The 5G Deployment Plan Handbook

The handbook for deploying wireless 5G systems with business case foundations.

Volume 1

Wade Sarver

The 5G Deployment Plan Handbook

Volume 1
5G technical deployment and history around building 5G and IOT businesses.

Wade Sarver

The 5G Deployment Plan Handbook

Copyright

First Edition © 2017 by Wade Sarver. All rights reserved. No part of this publication may be reproduced, stored in a retrieval system, or transmitted in any form or by any means, electronic, mechanical, photocopying, recording, scanning, or otherwise, except as permitted under Sections 107 or 108 of the 1976 United States Copyright Act, without the prior written permission of the author.

I am not a lawyer or an active certified safety expert. This book is completed based on research and my experiences. Safety processes and procedures are constantly updated and improved over time. The material contained is for reference only and may include products, information, or services by third parties. I do not assume responsibility for any third-party material referenced in this book.

This document is a guide to help people and not a guarantee that you will do everything properly. By reading this, you agree that myself and my company is not responsible for the success or failure of your business decisions relating to the information presented in this guide.

www.wade4wireless.com
Cover and design by Wade Sarver

The 5G Deployment Plan Handbook

Contents

The 5G Deployment Plan Handbook! 0
    The handbook for deploying wireless 5G systems with business case foundations. 0
Contents 2
Figures 6
Tables 6
Who is this book for? 7
How to use this Book: 8
Introduction 9
Why 5G? 10
    What is 4G? 10
        A quick history lesson. 10
    The 4G network 10
What is 5G? 11
    Quick history recap: 11
Will 5G replace LTE? 15
What Applications will 5G have? 15
What will the 5G be used for? 16
Why the Need for Speed? 18
Why the need for 5G Low Latency? 19
Why Narrow Bandwidth systems in 5G? 19
5G Network Slicing 20
4G and 5G Spectrum and Technologies 24
    4G spectrum, soon to be part of 5G Spectrum 24
    TDD and FDD Formats 25
The Wireless Network outline. 27
    The Evolved Core. 27
    The RAN 28
Wireless Deployment Planning Overview 29
        Pre-deployment Planning Overview 29

The 5G Deployment Plan Handbook

- Planning and budgeting for deployment. 31
- Start with the end in mind. 32
- Put some thought into whom you are going to serve. 32
- What is the service? 33
- Break it down even more. 33

Inter-Network Connectivity 34

RAN Backhaul and Fronthaul Overview 34

- Fiber connections: 36
- When is fiber used/not used? 37
- Microwave Connections: 37
- When is Microwave used/not used? 39
- What is LTE UE backhaul? 39
- Resources: 41

RAN Site Components 41

- BBU 42
- Radio. 42
- Antennas and Jumpers. 42
- The Mounting Structure and Hardware. 42
- Battery backup. 43

Testing at the site for more than the equipment! 43

The 4G deployment plan 43

- Types of Cell Sites 43
- The BTS Installation. 45
- The Radio Head Installation 45
- Antenna Notes 46

LTE MIMO Deployment Notes 46

From 4G to 5G. 48

The 4G and 5G HetNet 50

What will 5G networks look like? 51

- System Outline 53

What is the 5G System Plan? 54

- What is the overall 5G plan? 57

The 5G Deployment Plan Handbook

- The 5G System 59
  - Standard System 59
  - Base Station 59
  - Antennas and Radio Heads 59
- How does MIMO work? 59
- Deploying 5G Small Cells 60
- Will 5G be a Success? 63
- The 5G HetNet 64
- The Cloud RAN 64
- What is Edge and FOG Computing? 65
- What is SDN and NFV? 65
- What about Wi-Fi? 67
  - Cheap and Dirty 68
  - Carrier Grade 68
- Who will win in 5G? 69
- The Real 5G Winners Will have VISION! 70
  - Resources: 71
- The 5G Business Case Foundation 73
- What is your Business Case for Wireless Coverage? 73
  - Medical and Health Care 73
  - Utilities 77
  - Transportation 80
    - Rail or Bus 80
    - Highway 82
    - County and City Transportation 84
    - Air Travel 85
  - Unmanned Vehicles 87
    - Drones/plane 87
    - Automobiles 88
    - Boats 90
  - Emergency Responders 91

- WISP ... 93
- Small Carrier ... 94
- IOT Systems ... 96
- Enterprise ... 98
- Business or Building Owner ... 100
- Building Maintenance ... 101
- Entertainment, Stadium, Large Venue ... 103
- Smart City ... 106
- Construction vehicles and sites ... 108
- Renewable Energy ... 110
- Gaming ... 111
- Other – what will your business plan look like? ... 112

IOT ... 112

What is NB IOT and how will we use it? ... 112
- What is NB-IOT? ... 113
- Resources: ... 114

Glossary - Naming Overview (Abbreviations and Acronyms) ... 115

A Note from Wade ... 123

Other Books by Wade ... 125

Extras ... 125
- More business plan sheets: ... 125
- Other – Write your business plan. ... 126

Scope of Work Outlines Cover Sheet ... 128

Scope of Work Details ... 129

Figures

Figure 1	11
Figure 2	14
Figure 3	15
Figure 4	18
Figure 5	21
Figure 6	22
Figure 7	26
Figure 8	28
Figure 9	29
Figure 10	32
Figure 11	34
Figure 12	35
Figure 13	49
Figure 14	51
Figure 15	53
Figure 16	54
Figure 17	55
Figure 18	58
Figure 19	64
Figure 20	64
Figure 21	116
Figure 22	116
Figure 23	117
Figure 24	119

Tables

Table 1	56
Table 2	56

Who is this book for?

For anyone who may be interested in deploying 5G systems. This book is concentrating on 5G system deployment ideas and ways that we could deploy. I am hoping to prepare you and your teams for the roll out of 5G systems. We often don't think about these things until it's time to deploy. Then we brainstorm and plot, spend long nights and weekends to come up with this conclusion, "Let's do it like we've always done it!". You know I am right. Very few people or companies take the time to understand how the systems should be deployed until they started deploying. I am no better, but I was hoping that this may help prepare you to improve on the past processes by going over what 5G is, who will use it, and how it will be deployed. The roll out will be something that will start slowly with some experimentation, probably by Verizon Wireless and AT&T Wireless because they have the deep pockets to get started. But don't you want to do something soon? Don't you want to be an influencer, someone who tried it, even on a very small scale, showing the big boys that even the little guys can do it? I thought so! That is why I took the blogs I wrote and added more information to prepare this book for you.

If you are reading this, you could be an installer or a commissioning engineer or RF design engineer or an optimization team. Anyone of these groups may find this book useful. I want to help others move ahead and learn from what I have learned. Feel free to contact me at wade4wireless@gmail.com to give me feedback. I have been putting information up about wireless deployment on my website, www.wade4wireless.com for all to see. I thought by putting the information out there then people would be interested in the small cell deployment. Feel free to sign up for my newsletter.

My goal was to inform you and help you along in your career. 5G and IOT are somewhat new to the wireless market, and this may help you enter the market quickly by taking what I have learned and applying it to your business.

How to use this Book:

When reading this book take into consideration that here are links to help you. I write this like my blog and podcast, giving you the opportunity to look up the information where I found it. I also give you links to opinions that are very different and new items. Things will change over time, so this is something that will help you understand where we are now and where we are going.

I explain things the best I can so you can understand, but I have a glossary in the end that will help you learn the acronyms and a resources section with more links for you to research and learn. Many of them have PDFs that you can download to look at the existing research. Many links are from other publications. I did the research, so you don't have to. What I did here is put the information together so that you can put it together and form your ideas about deployment. Which way is best and what will happen? Let's journey together into the 5G realm.

When reading this, take the time to look at the options and different types of protocols and deployments, then decide what you need to do for your purposes. Not everyone will be deploying carrier grade systems; some will be license free or enterprise systems or very small arena systems that they need to figure it out for themselves. Here you need to take what you need and use it for your purposes to make the best network you can within your budgets.

Learn what you can and continue to research when you can. I put this together to help you get all the information in one place where you may take weeks to find it yourself. Helping you get started and focused on what you need to do and learn what you can about 5G deployments.

The 5G Deployment Plan Handbook

Introduction

Here is another installment of the Wireless Deployment Handbook (WDH) series. Many people ask me about 4G and 5G and how we get there from here. Well, since you asked I thought I would cover the deployment part of the networks. Are you interested in how to deploy and how the transition happens?

You all know that 5G is coming and it is going to open the doors to new applications such as wireless virtual reality, new IOT ideas, Artificial Intelligence applications, low-latency offerings, wireless cloud computing, and so much more. It's a game changer if it's done right.

Will the network be more than a cool this? It's becoming a necessity for any of us to do business. The broadband connections need to be made available anywhere making this the new utility. We already see it with the mandates coming from the US Government that broadband is a necessity for growth and business. We all need electricity in our homes, right, now let's add broadband. The telephone has been replaced with the broadband connection. The difference is that the phone was in every home, now the phone and a broadband connection is with you every minute of the day, by your side as a smartphone, tablet, or laptop. We all feel the need to be connected most of the time.

This constant connection is such an addiction in most countries that we see addictions to being connected. The need to stay in touch has totally overwhelmed some people, even me. I now know that I can manage my priorities in life.

Staying connected can be a good thing. Look how families can live remote yet communicate every day for pennies. They don't have to talk, but they can text and take pictures. Remote family members can see their relatives, children, grandchildren, nieces, nephews, aunts, uncles, brothers, and sisters daily and even over holidays without taking the time or expense to visit. We can talk anytime and share those special moments with all of them with the click of a button on your smartphone. How cool is that? See, technology is a good thing when properly managed.

So, let's dive into the networks that make all of this happen. Let's look at the new generation, 5G, and how to deploy it so that we don't let these families down. Let's support them and increase the functions and ways to stay in touch. Let's make all of this happen. It is up to you, the deployment teams, the unsung heroes that make this happen with little or no recognition from anyone, even the customer who wanted this system. Sure, they will take most of the credit, but you have the bragging rights knowing that they could not have done it without you! Congratulations on taking the first step her, now, today!

Why 5G?

Right now, it's a buzz word, or number, or letter. Anyway, the 5th generation of wireless systems, specifically carrier's wireless system. I go into the brief history below, but 5G is going to be a compilation of several networks.

What is 4G?

A quick history lesson.

We all know about 4G and what it is, right? It is the network that replaced 3G. What is 3G? Well, 3G was mostly voice and data, but it was very limited on data and was geared to provide reliable voice. The protocols used were CDMA and GSM around the world for carriers. They worked very well and were built out is many bands of spectrum around the world.

You may have known all of this, but I think it is good to put this into perspective so that you know the evolution of the network. It didn't just happen for no reason; the end user consumer drove the market and the network. Carriers had to keep up, and they decided on LTE as a common standard.

With 3G there were two main standards, GSM and CDMA, which drove competition and made devices work only on one or the other. If you had a GSM phone, it may work in some countries with some carriers but not with others. If you had CDMA, the same might be true. What did the carriers do? They found a way to make the format a commodity by all settling on LTE. Now any device can be used on any LTE system assuming the device has the spectrum availability in it.

Spectrum is the band, the frequency that the phone communicates on. Now almost all the devices have every band available in them. In the USA, you can take your iPhone and open it up to work on any major carrier's system.

The carriers made the format a commodity; they also made it easier than ever for the consumer to move from one carrier to another making competition fiercer than ever. It was not exactly a win-win for them. They made the format a common format; they made it so that their cores can be centralized and on the cloud, cool. We will get into that later. For now, we will look at LTE as the 4G format.

The 4G network.

We should be looking at the network, specifically 4G. It is the foundation for the future networks. Mostly due to the carriers' budgets. They had to do a forklift upgrade to go to 4G. What that means is that they had to install all new equipment to go to 4G because they plan to sunset the 3G system. When they did this, they realized that they couldn't keep putting in new networks. One of the reasons they went with LTE was because of its versatility and the new options that can make the end user see upgrades on one format. It is cool.

To name a few, which I will get into later. The one thing that you need to understand that it is an all-IP network, which means it is a complete data system from end to end.

End to end means that from the network or even your laptop it goes through the network in an IP format and when it goes over the wireless network it still is pretty much an IP packet traveling all the way to your smartphone where it remains an IP network.

The 5G Deployment Plan Handbook

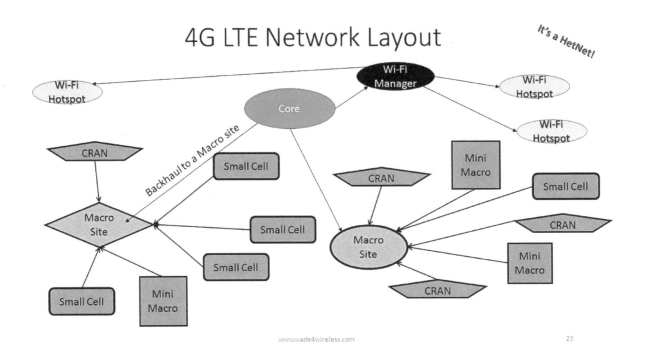

Figure 1

What is 5G?

The 5G network, as of 2016, is still being defined. What we do know is that it will not be like 2G, 3G, or 4G because it will be more than the format, spectrum, speed, or even the equipment. Let me break it down for you.

Can we define 5G? Let's look at the Wikipedia definition of 5G, found here, "5G (5th generation mobile networks or 5th generation wireless systems) denotes the proposed next major phase of mobile telecommunications standards beyond the current 4G/IMT-Advanced standards. 5G planning includes Internet connection speeds faster than current 4G, and other improvements." What does that mean? It's more than just a network connection or a format. It will include the connection to the internet, the connections to each device, the broad-spectrum of devices used in the network.

Quick history recap:

- Older formats were defined by what they could do, but we just looked at the wireless format. We looked at 3G as GSM or CDMA. We were looking at 4G as the next generation which was LTE as chosen by the carriers because it is the Long-Term Evolution of wireless.

- Then when going to 4G it was a competition between WCDMA and WiMAX and LTE, LTE clearly won the battle. All the carriers went with LTE helping them make the equipment and deployments more of a commodity which saves them money. The evolved packet core made it easier to distribute the radios and split up the core. The all-IP system matched what most networks are today making the transfer of data more efficient and clean.

- So why improve? Because we're human, that's what we do, advance. In this case, it was the end user's insatiable demand for data that has pushed out 3G quick, costing network operators a lot of money in upgrades to get to an all-IP LTE system. Thanks to the iPhone, the mobile device changed forever!

- The big difference? It's the network! Going to 5G is more than just the wireless format, it's all about the network and the combination of networks. Back when 4G was coming out there was this concept, the HetNet. The Heterogeneous Network is a concept that came from the computer world where, per Wikipedia, *"using different access technologies. For example, a wireless network which provides a service through a wireless LAN and is able to maintain the service when switching to a cellular network is called a wireless heterogeneous network"*.

The HetNet is the game changer along with new speeds and spectrum and formats. When looking at the system, you could have macro sites and small cells, LTE and Wi-Fi and perhaps another format all working together as one big happy network where the end user has no idea what network they are on. You could be in any spectrum, 600MHz, 700MHz, 1.9GHz, 2.5GHz, 24GHz, 28GHz, 60GHz or another band which could be allocated to 5G. You could even be in the unlicensed spectrum running Wi-Fi or LTE-U or a lightly licensed band like 3.65GHz, the CBRS, here in the states. The end user may notice the change in speed but not the format or spectrum change. In fact, I would believe the end user won't care unless they see a big change is speed, or quality of experience, (QoE). Seriously, do you even think about it unless voice is crappy or the download rate is painfully slow, or you lose connection altogether?

So, what is 5G? It's the combination all the network encompasses. It will be all the parts put together to make the speeds super-fast. Now, you're probably wondering how we will get there. Lucky I put together this list for you to see how we will improve speeds.

- Carrier aggregation – what this is the method used to aggregate carriers, which is explained by 3GPP here. What that means is that carriers used now can be combined in the equipment to look like one big pipe of bandwidth. It is advancing. Currently, I have seen three carriers all put together, but it should grow to 6 or 7 soon allowing the pipes to be bigger.

- Carrier aggregation with unlicensed bands – I thought I would throw this in there because it is very different that normal carrier aggregation. I will tell you why! Licensed aggregation is from the same BTS making it easy to aggregate, but the unlicensed aggregation like LAA and LWA is combining spectrum from a BTS and some unlicensed access point. That makes it much more complicated, and I must give the OEMs so much credit to do this. In the UE device, they can put it together, and it seems that Qualcomm figured out how to do it on the device.

- Massive MIMO – that's right, the antennas are making a difference. I know, it's more than the antenna but let's just point out that it's a team effort between the radio and the antenna to shove even more bit per second in the same bandwidth. There is a high-level overview here. I am not going to get into the technical details but the beam forming technology and the way that one antenna will have hundreds of antennas in it that can focus on one user, is amazing. I remember that Ruckus has high-tech antenna technology in the Wi-Fi spectrum which set them apart from their competition. The antennas will push data to new limits in 5G systems.

- Improvements in LTE – the formats are improving but bandwidth is limited in today's spectrum, so this is reaching its limit. However, we now have LTE-Advanced, released in networks in 2016. LTE-A includes much of the services that are listed here. However, the radios need to improve, or we don't advance.

- New spectrum – the spectrum is coming in bigger bandwidths for the carriers to put together. We no longer see carrier use 1MHz carriers, but they are looking for 5, 10, and 20MHz carriers. When the "5G" spectrum in the mmwave, (millimeter wave), is released they will have 20 MHz channels and higher. Imagine a carrier has 100MHz of bandwidth on one carrier, and they can dedicate that to a limited number of users, and they can aggregate it with three other 100MHz wide carriers to provide 400Mhz of spectrum. Would that compete with cable for home internet access? I think so, as a fixed wireless system where we no longer must run cables or fiber to a house or business. If only the carriers would work out a flat-fee unlimited data plan for users that would rival the cable companies plans without the TV channels.

Now, I went over the wireless improvements, but as you know, it's more about the network which includes the backhaul and core. Did I say backhaul and core? You know it's more than that!

- SDN – Software Defined Networking which makes the routing architecture smarter and more efficient. If you want to learn more, start here.

- NFV – Network Function Virtualization used SDN to make the network virtual. That will make the network functions work closer to the user. Learn more here.

- Cloud Computing – here is where the applications are brought closer to the user, lower latency and improved customer experience to the point where the network sees less congestion. Learn more here. So where is the cloud? It should be on a server near you. They could be anywhere in a remote data room near you to lower latency.

- Edge Computing – taking the applications beyond the cloud to get it running on a server very close to the end user. For more go here. Also, called Fog Computing, see next item.

- Fog Computing – (a term made up by Cisco) this is taking the cloud and shoving it as close to the end user as possible, to the edge where the IOT will be able to make smart decisions in very little time, low latency. I found a good explanation here.

- Cloud RAN – C-RAN is where the RAN will not have a local BBU, but a virtual BBU. Like CRAN which is Centralized RAN which is where the BBU hotel is remote, and fiber connected the BBU to each radio head which could be using CPRI or another format. The limitation with this is that the fiber needs to be dedicated fiber for each radio head. I have an article here, but I want you to realize that if you are in the industry, then CRAN and C-RAN are very different, ask any OEM or carrier. Cloud RAN is where the BBU function is more virtual whereas Centralized RAN has a direct physical connection to the BBU. Get it?

5G will encompass applications, new ways to use the Het Net. New ways to get the processing power to the edge of the network using the cloud and even fog technologies. I think that we should change the paradigm of the wireless network. It won't be long until we have fixed wireless providing internet access to homes to replace the cable modems we need now. Operators will have more than 10Mbps backhaul for wireless cells, small or macro. It won't be long until they need 1Gbps to satisfy the needs of the end user.

So now we have 5G to be more than just a new format or higher speeds. It will be a combination of formats with so much more included. We will see 5G specific applications that will shape the network. We will see the networking equipment be a requirement, the cloud; even fog computing will be part of all of this.

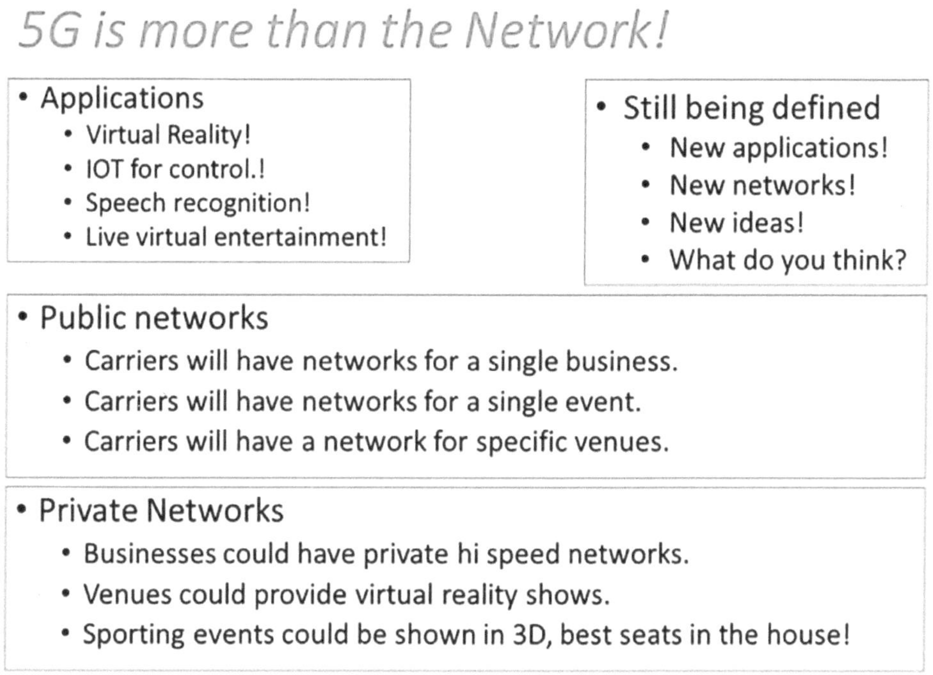

5G is more than the Network!

- Applications
    - Virtual Reality!
    - IOT for control.!
    - Speech recognition!
    - Live virtual entertainment!

- Still being defined
    - New applications!
    - New networks!
    - New ideas!
    - What do you think?

- Public networks
    - Carriers will have networks for a single business.
    - Carriers will have networks for a single event.
    - Carriers will have a network for specific venues.

- Private Networks
    - Businesses could have private hi speed networks.
    - Venues could provide virtual reality shows.
    - Sporting events could be shown in 3D, best seats in the house!

Figure 2

Will 5G replace LTE?

No, LTE will be the foundation for the 5G network. LTE will get new terms, like LTE-Advanced, just like 4G will become 4.5G and 4.9G before 5G is released. You will see great increases in speed and LTE will become more and more advanced. It is quite impressive what the OEMs are doing.

The HetNet will be an applicable term. Networks will handoff between so many different types of systems. However, I do believe that the 5G systems may be a type of LTE to hand off to each other properly. Remember that there will be multiple systems running simultaneously. They would need to have a seamless handoff for the best user experience.

I will use an example. The carriers have truly adopted wi-fi because of the complexity of the shift from LTE to Wi-Fi, that is why the carrier are so excited about LTE-U. The see a viable way to make the best use of unlicensed spectrum. The Wi-Fi carriers don't like it because they invested a lot of money into their license free Wi-Fi networks.

The 5G systems will need to have a way to such seamless handoffs without dropping a call or a session.

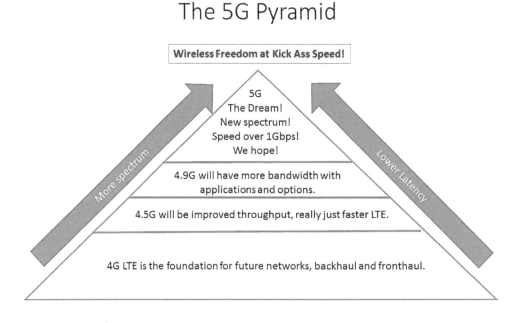

Figure 3

What Applications will 5G have?

For one it will have all that you do now on 4G, internet connection, all the apps, all the things you're doing now that you feel you can't live without.

The new applications will push the network beyond the limitations that we know today and into virtual reality and IOT connections and more streaming video that we could not have before.

One more thing that is driving it? Vehicle to Vehicle communications. It is the thing that we expect to change the way we live. Vehicles that can communicate with each other to make the chances of accidents lower than ever, in theory anyway. It is also pushing the limits of driverless cars. We are hoping that someday in the next five years that driverless cars are commonplace. Can you imagine that? It would reduce the chance of death on the highways and roads in general. The network will be responsible for all of this, albeit more than 5G but the network reliability, latency, and speed. This falls under the Internet of Things, IOT.

In the enterprise, we will have real-time reporting of KPIs, or stock trades, or horse races that we can get real-time results on, even see the action take place in real time. '

Sporting events can offer you the best seat in the house at your home, or at a bar or even at any remote location by showing the game in virtual reality. Think about that, watching the game, be it American Football, Soccer or football, baseball, rugby, cricket, or any Olympic event as if you were in the stadium. The only thing you won't get is the smell of the arena or someone spilling beer on you. Invite me over, and I can take care of the beer spilling part.

What about smart cities? Suddenly we pushed cities into the idea that they can see all the city at any time. I know you're thinking that big brother is watching, but what if big brother was looking at traffic patterns, accidents, traffic delays? They may be able to help or at least report it so that you know to go a different way, and in real time. They would also see major potholes in the road and warn is that it is ahead.

The smart home will go to the next level. When the new networks come online, we should have improved battery life with greater efficiency so that we can take the devices out of the home, away from power, and rely on batteries for months instead of days. We could track pets, bikes, anything that we leave outside for a fraction of the cost it would take to do it now.

Health services for taking medicine and tracking health conditions will improve, we see it now, it will go to the next lever for real-time reporting anywhere and anytime.

Drones are always brought up, but the network for drones should be large. With drones, they may have a near field 5G connection that would allow them to control and upload and download data. It is the network that will enable these devices to get their information, but they will need to be autonomous at some point. The network can give them the updates and information they need for the flight path but they need to be able to talk to each other in the air, this is where a small 5G mm-wave network can come in handy.

What will the 5G be used for?

Fair question, but most of you reading this understand what is already on a wireless network. The wireless network carries a lot of traffic. This time when we look at 5G, we will intend to lose site of the wired connection. I would bet you already have lost the wired connection in your home. Let's face it, at home, most people rely on Wi-Fi, they hook the router into the internet connection which could be a cable modem, a fiber router, or even a carrier's hub which receives LTE but connects to your devices via Wi-Fi. We all love Wi-Fi which is part of the 5G network as part of the HetNet.

The 5G Deployment Plan Handbook

That's right; I explain what a HetNet is but all you need to know that 5G is not one network or a specific format. It will be a collection of many networks and formats to make one big IP network.

What will be on it? Let's look below:

- End users and uses
    - Tablets
    - Smartphones
    - Laptops
    - Watches
    - Other wearable devices
    - IOT Devices
    - Cars
    - Drones
    - Home devices
    - Video cameras
    - Body cameras, (Law Enforcement or Fire Rescue)
    - Smart cities – traffic lights, traffic monitors, light control
    - Utility meters – electric, water, gas
    - Gauges for heating and cooling control
    - Home Internet access, replacing cable modems!
    - Video gaming
    - Virtual reality
    - So much more to come!

That is what I have seen so far, and the list keeps growing. We have relied on wireless for years, and now it's no longer a luxury but an expectation that we have wireless everywhere. We expect great coverage and signal everywhere and yet there are so many protests against new towers. While towers are not popular, although I don't know why because I think they are the coolest things to look at, carriers and deployment teams had to get more and more creative. Small cells and oDAS have taken off. It has cost the carriers more money and yet the end user expects to get lower monthly rates. I can't blame them, I don't want to pay more and when you see what they make in profit and what the CEOs get paid, it is hard to feel sorry for them.

If you are in this business and you deploy networks, please don't just look at it as a job but you are driving the progress of this industry in your country and the world.

Let me say it like this; the installers are putting in a new system, and they look at it as a regular installation. While they look at it as something simple, it is a foundation to making the system work properly. It's a shame the carriers look at this as something of a commodity. Installation and design can make or break a successful system. By successful I mean one that has reliable coverage or one that drops calls or data making the user experience suck. That's right, a bad user experience sucks. Did you ever drop a call, lose connection, or just plain have a crappy experience? I have, and it sucks because then you spend the next 5 or 10 minutes trying to call back or get a new connection. Frustrating! Whom do you blame? I blame the carrier. I know there are hundreds of reasons why the coverage is bad, walls, dead spot, overloaded site, but all the same if I know that then the carrier should know that and do something about it. Especially in a train station or an airport where people rely on their devices, just improve the experience already, will you!

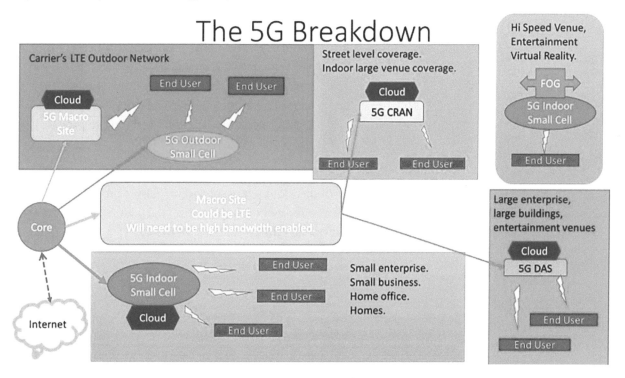

Figure 4

Why the Need for Speed?

Why do we keep improving the network speeds, the coverage, and the equipment? Because the customers are demanding it! They wanted this for years. You're thinking just the millennials are pounding away on their smartphones, right? It's way more than that! So many businesses use applications, enterprises that want to do more. Entertainment industries want to do more. Utilities want to do more. Transportation and governments want to do more, and they want to do it everywhere. The world is more and more connected thanks to wireless connectivity. We can connect anywhere there is coverage either on a terrestrial network or a satellite network. We live in amazing times. You must be excited to see the technology.

Some of you think we've gone far enough, don't you? We have not and we will continue to push our networks forward. That is what we do; we go to the moon then talk about going to Mars. On this planet, we want to do more, and we learn from the past. We surpass all our expectations. We want cars that drive us around. We want things that talk to each other without human intervention. We want to get on the Internet anywhere to catch sports scores or do a web search. We need all of this.

Why do we need 5G? Well, the iPhone changed the world and the dynamic. You see, that device made customers realize how much they needed the internet and younger people preferred to texting of talking. You could text in 3G, not a big deal but the internet access to your mobile device changed everything. The new thing that came out was the app, the application you could run on your phone. On the computer, they called it a program, but apps became mainstream with smartphones. Smartphones are the connector to most people and the internet, and they are very mobile.

Why the need for 5G Low Latency?

The key to true 5G high bandwidth needs as well as low bandwidth needs. The quick response for most devices will be needed so that applications can "talk" as close to real time as possible. You may have seen the RTC, Real Time Communication, a term tossed around RTC is where the device needs to react very quickly, and there is little time for delay. For instance, self-driving cars., They must process the data, so when they communicate with the devices around them, they need to have as little delay as possible. I am talking microseconds, not milliseconds. Why? Because they still need time to process the data.

Self-driving cars won't just talk to the network, but they will be talking to cars around them, "looking" all around them, driving the car, making thousands of decisions every second, millions every minute. Deciding how to prepare for the road ahead, the environment around them, and what's the next move. They will always be concerned about what's next outside the car and inside the car.

Therefore, the communications system must talk quickly, hence, low latency.

Why Narrow Bandwidth systems in 5G?

Narrow band is for IOT devices. You see, with LTE and Wi-Fi, they tend to be on the air all the time which means the device, a smartphone or your laptop, will be listening and processing data all the time. With IOT devices, they don't need to talk all the time. They could be pinged once a day or even just talk when they have something to say.

While there are several reasons, the main one is battery life. If it is talking all the time, then the power draw is constant and high. Broadband kills any battery because it is talking all the time. To get a 10-year battery life, you need to plan when or how it will talk and listen. You don't want it drawing on that battery 24/7 because it's listening and processing data. Think about your laptop and how the Wi-Fi will drain the battery life, just like the display. These are the main draws of power. Well, with many IOT devices there is no display, so they only massive power draw is the radio. If you can have the radio go to sleep until it is needed or to wake up at a time of day, then the battery will last a very long time.

For example, if you have a water sensor or a gas meter or a water meter, three devices that could be mounted where there is no available power source, you need to make sure that battery will last a very long time. Each device will have a different function.

- The water sensor may only wake up to send a beacon to let the system know that it is alive working unless there is a high-water alarm, then it will send out alerts. This way the battery will only work when it must.
- For the metering, gas or water, it doesn't need to send information all the time. Only maybe once a month or when it's queried. It may send information of the usage is extremely high to let people know that there is a massive draw on the product measured. This way the battery will last a very long time and the company deploying these devices will not need to run power to everything.

These are just a few examples of how the narrowband will be a slice of the 5G network.

5G Network Slicing

Network slicing is 5G's way to get you everything. You see, one network will not provide all services for everyone, so they have 5G which will encompass many networks, wireless networks, into one big network. You can't do everything with one wireless network. Like Steven Wright says, "You can't have everything. Where would you put it?" If you had one network, it would not be efficient enough to serve all the devices on it. You want a network that works. Otherwise, you have a *notwork* because it does not work! Most IOT devices don't need broadband. Most smartphones need mobile coverage. Most laptops need broadband. Most gamers need massive broadband to get the VR to work. Each specific group has a different need. Wouldn't it be nice if you could have several different wireless networks and have them all go into one core and share resources? Well, 5G came up with network slicing so we can do just that!

The research on network slicing showed me one thing that this is a fancy way to say different networks all connected to a common core. I think this term is interesting, but if you are in IT, then you know that you could have multiple networks, virtual or separated, all sharing the same backbone or even the same physical network. The way I see it, it is all about the RAN! Let's explore why.

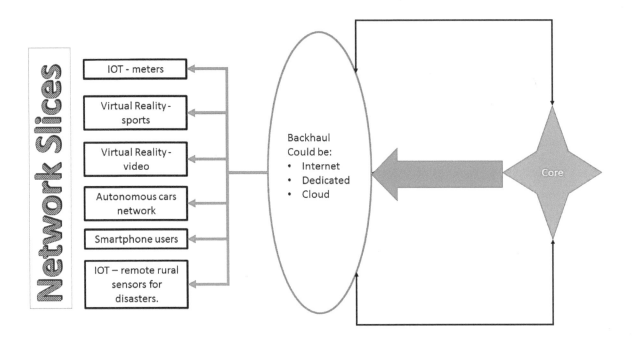

Figure 5

Well, in 5G, it is not much different. The big difference is that you could have a wireless network dedicated to a specific service. What this means is that when planning a network, in this case, a RAN network, make sure you know what the application will be so that you can plan accordingly.

Think about the different markets 5G will be serving. It could be autonomous cars, virtual reality, or tons of simple IOT devices. Each system will have different need and purpose. The goals are not the same for each. Therefore, they should not all share the same network. So, for the 5G network to include them all, they came up with a cool term like network slicing. The reality is that they will all be different networks that could be sharing the same core or even backhaul. We are creating a way to share resources and build in efficiencies.

We'll get into why in a few minutes, let's look at how they will work together first. It's all about sharing of resources. Think of the HetNet, (Heterogeneous Network) and how we had small cells working with Macrocells and Wi-Fi all working together as one network. Now you have multiple networks all working independently, yet, connecting to the common core.

Which resources are shared in network slicing? The backhaul and the core but also routers and servers and possibly even cloud resources. The key to getting latency down is to rely on the cloud. However, the end use will determine which network will be used and how it will be utilized. The way I see it, from a wireless viewpoint is that the device will need to have a wireless network that fits the needs. In other words, virtual reality with need low latency and very high bandwidth to work properly. Autonomous cars will have very low latency but lower bandwidth needs. IOT devices will have medium latency but very low data rates, and they will not be listening to the network all the time like the other 2, they will only listen to the network on a need to know basis.

The examples above show us that there will be a need for specific wireless networks to serve each purpose. The common denominator will the core. The core will need to know how to process each part of the network. Making the major carriers happy that they have resource sharing capabilities to save costs. They want to reuse as many resources as possible. Device manufacturers will continue to improve devices and battery life.

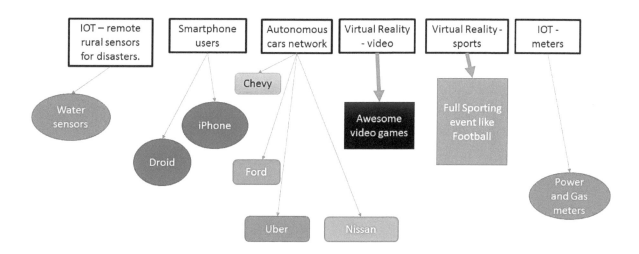

Figure 6

Although, battery life is still an issue. While battery life has greatly improved, the power draw is so much higher than it was five years ago, Hell, it's so much higher than even a year ago, While the processors are drawing less and less power, we have higher demands on many of our devices, like the smartphone. We want bigger and brighter displays, and we are on them for most of the day not only to talk but to gather data. Even when you are not talking on your phone, the chances are good that it's getting updates for email or other data without you even looking at it, drawing on the battery even more. Not only that but the constant communication with the LTE and Wi-Fi networks are drawing power all day.

Back to network slicing. We will have several different use cases for the network, which will require a specific last mile network to serve the purpose. It seems a bit crazy to have multiple wireless networks until you realize that billions of devices will be connected and each group will have a specific purpose. Each group will have a unique revenue stream. Some will be high usage and draw more money per month and others will have extremely light usage and will only cost pennies a month. Each slice of the network is built for a specific purpose, and the billing for each slice will be dramatically different. Here are the efficiencies.

These networks will be running in parallel to each other. They will be independent of each other but have a common core. With the growth of software defines networking, SDN, and Network Function Virtualization, NFV, the networks will become smarter and smarter and start to improve efficiencies without human intervention. It's already happening, but it will get better and better with improved efficiencies.

While all of this will be interconnected, they will be isolated from each other. Some networks will be independent of each other. The key to slicing is even though networks share resources, they will not be reliant on each other to keep the network up and running, (unless the core crashes).

The drawback is the core will control everything. Get to the core, and you get to the heart of multiple networks all at one time. If they make changes to the core, they need to be sure it will not affect the other networks. I would imagine that updating the server controlling the IOT network would have no effect on the autonomous driving network. But, what if it does? Then a real problem will be at hand!

The core will be the key connecting point to these networks. Running on the cloud should help efficiencies along with the rise of the virtual core, the impact should be minimal. Just remember, they all need a brain, and that brain is the core!

You could have several companies serving several markets, like the carriers taking care of smartphone users and someone like SigFox working with the IOT users and maybe someone else taking care of virtual reality and yet another company taking care of autonomous automobiles.

While this is a slice if heaven, (sorry, I couldn't resist), we expect each slice to be running independently of the other even though they share a common core.

Resource:
http://www.5gamericas.org/files/3214/7975/0104/5G_Americas_Network_Slicing_11.21_Final.pdf

4G and 5G Spectrum and Technologies

4G spectrum, soon to be part of 5G Spectrum

The spectrum is whatever they could get from the FCC in the USA. They get it from the spectrum auctions that the FCC holds. There is always a need for more although some carriers have yet to deploy all of what they have. With 3G they could use smaller swaths of bandwidth. 4G changed that, and 5G will only make them want more.

Spectrum is tough to show because there is 4G spectrum for auction here in the USA. I realize that spectrum goes to the highest bidder, (in my opinion small businesses suffer). However, the rush to get spectrum has diminished by the carriers learning to make the most of the existing spectrum. While the bands are small, they have been using something called carrier aggregation to combine spectrum bands to look like one big pipe, which is awesome. The OEMs have worked to put together 2 or more bands so that they look like one big band making the end user happy with more throughput.

In the USA, there are many bands.

- 710 to 716MHz paired with 740 to 746MHz used by AT&T
- 746 to 757MHz paired with 776MHz to 787MHz used by Verizon Wireless
- 806 to 866MHz and 869MHz which belongs to Sprint, this is the old Nextel band.
- 1710 to 1785MHz and 1805 to 1880MHz is T-Mobile AWS spectrum.
- 1850 to 1990 MHz is Sprint FDD spectrum.
- 2.5GHz to 2.7GHz is Sprint TDD spectrum.
- More and more, it would take some time to break them all out, so I am providing you with some websites to get all your spectrum information. So much spectrum is out there, and the carriers are grabbing what they can.
    - https://en.wikipedia.org/wiki/Cellular_frequencies
    - https://transition.fcc.gov/oet/spectrum/table/fcctable.pdf
    - http://www.ntia.doc.gov/files/ntia/publications/2003-allochrt.pdf
    - https://www.qrctech.com/v/Assets/Pageimgs/FreqChart.pdf
    - http://specmap.sequence-omega.net/
    - http://www.phonescoop.com/articles/article.php?a=99&p=1493
    - http://www.ntia.doc.gov/files/ntia/publications/spectrum_use_summary_master-07142014.pdf
    - http://www.arrl.org/frequency-allocations

- http://www.phonearena.com/news/Cheat-sheet-which-4G-LTE-bands-do-AT-T-Verizon-T-Mobile-and-Sprint-use-in-the-USA_id77933
- http://www.droid-life.com/2015/02/05/us-wireless-carrier-bands-gsm-cdma-wcdma-lte-verizon-att-sprint-tmobile/

TDD and FDD Formats

There are two technologies for LTE. For LTE, they have FDD and TDD which both are viable options. Both are viable options. They are both used by carriers in the USA although FDD has been the choice in the past.

- What is FDD? FDD – Frequency Division Duplex is something that was used commonly in 3G. It's paired spectrum with an uplink band and a downlink band in their specific spectrum. For 1G, 2G, and 3G this was common so you could have a talk and receive channel in the system. There is a guard band in between the transmit band and the receive band. FDD was very popular with GSM and CDMA. It is very difficult to take advantage of MIMO antenna technology in FDD compared to TDD.

- What is TDD? TDD – Time Division Duplex is where there is one large piece if spectrum used for uplink or downlink. Any part or percentage can be assigned to be the uplink or downlink. If you have 20MHz of bandwidth available, then you're not locked into 10MHz up and 10MHz down like FDD. Instead, you have full control over how much goes up and comes down. The downside that some carriers had was the timing of the spectrum, and it's higher bands that have this. However, Wi-Fi spectrum is pretty much all TDD, and it works quite well for data. On the other hand, WiMAX used TDD, and it seemed to be taking off but it never fully blossomed and was cast aside for LTE. TDD makes MIMO technology easier to use because it is all in one band.

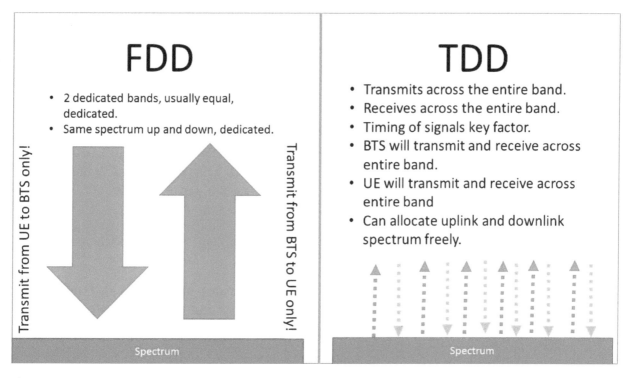

Figure 7

So, what can LTE do? It can do both, and it does do both. Just not the same equipment. You could have equipment do either LTE-TDD and LTE-FDD. Both are released commercially as well as part of the 3GPP standard. When you look at the deployments, it helps to know which format will be deployed. You see, FDD may need two antennas or a combiner to work on a tower. While TDD is all in the same spectrum and the same antenna is used for transmit and receive. The way that today's radio heads work it isn't much of an issue anymore because they can handle the formats quite well. In 2016, you still can't run them together in the same radio head, although the OEMs are working towards that functionality. Antennas are being designed to run both together by adding more ports and more weight to the antennas.

Note that Wi-Fi is TDD and ZigBee is TDD. Most Bluetooth is TDD. TDD appears to be the choice moving forward. Most 2G and 3G systems were FDD, and they are being phased out.

Carriers are learning that when everything becomes truly digital in IP format that it will matter less and less for the BTS, but antennas and spectrum efficiency become more important. As of 2016, most of the carriers already have implemented VoLTE into their main networks, all except maybe Sprint who was still relying on CDMA to carry the voice. The carriers know that when they convert VoLTE, it should be the last step to dismantling the 3G networks, saving them money in the long run by retiring 2G and 3G systems.

Sprint is a great example of having both. They have FDD on the CDMA and LTE carriers as well as TDD for their 2.5GHz spectrum of LTE. They have a huge amount of spectrum in 2.5GHz, over 100MHz of bandwidth. They have successfully deployed on both formats.

I found a great resource at http://electronicdesign.com/communications/what-s-difference-between-fdd-and-tdd at Electronic Design, an article written by Lou Frenzel.

The Wireless Network outline.

The network will look like what we see below. Let's point out that the network is more than the equipment at the tower site or in the tower. The equipment does go back to a core network like the 3G system did. The difference is that now the core can be almost anywhere.

The Evolved Core.

There could be a central core for all the sites, or they could have smaller cores that all talk on an IP backbone. The pros of a central core could be cost, on location, but the issue is that you should connect to every site out there, so you need to think of backhaul costs. These costs add up and not just for the upfront engineering costs but the monthly cost for each link. Most of these may be fiber.

Small cores have efficiencies on backhaul connections. All the cores need to speak to each other, but this would be on a large IP backhaul with plenty of bandwidth.

The wild card that we have here that we didn't' have on the 3G network is the cloud. Cloud computing moves much of the central core functions away from the core. As a matter of fact, the RAN may soon be just a radio head with a router. The core is becoming virtual. The cloud has changed much of this.

With the upgrades in network and computing, we have several options. NFV, network function virtualization, and SDN, software defined networking, have revolutionized the way IP traffic gets routed. It helps improve efficiencies in the network and provide alternate routes for traffic that may have been very hard to set up previously. The backhaul network is becoming smarter and more reliable.

With this, computing isn't just done ion one server anymore. The services that once had to run in one location are now being pushed out closer to the end user. The key is to provide servers that can extend the processing power of the central server.

You have seen this on your computer where you run an application, app, from the Internet and most of the computing is very quick with very low latency.

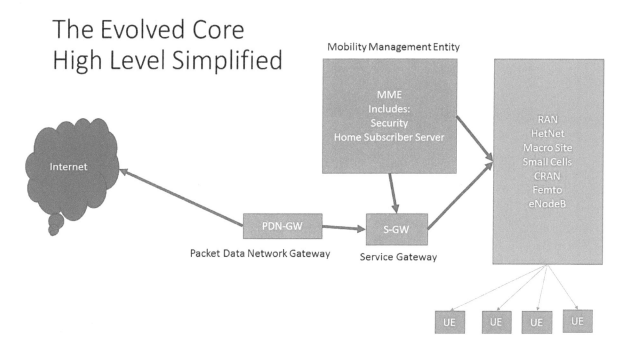

Figure 8

The RAN

Here's where we all look at the radio network. It is the most expensive part of the network. It will have many parts and pieces and could incorporate even more parts as the network matures. Let's start by explaining that **RAN means "Radio Access Network,"** and it will have everything outside between the core and the end user. Most people just think of the radios, but the network is more than just radios and core. It is a complex system of connections that need to talk to each other and the core and the user's equipment, the UE. The UE could be a smartphone, a laptop, a device in a meter or a video camera, or anything that can connect to the network. Don't limit yourself to thinking it is just LTE because it could be Wi-Fi or another type of wireless format. 4G is a collection of high-speed formats and 5G will only add more formats and complexity to the network. It's something that you need to be aware of when moving ahead. Although Wi-Fi never panned out as the carriers had hoped, it is still a major part of the network for offload.

Remember that this book is about deployments. We're not diving too deep into the architecture. The heart of the RAN is the BTS, base transceiver station. The radio itself. The eNodeB is much more advanced that the radios of old. It could be any spectrum, but to give you an idea of what is in it I made a drawing below that is typical of today's BTS.

Remember that there is more to the BTS than just receive or transmitter. It is also a router that connects the backhaul which could have microwave. The BTS also has batteries to survive outages. Power backup will be in most macro, and small cells Wi-Fi usually won't have power backup. Now that we have 5G you will also see servers at more sites to support cloud and edge computing. We need the radio heads at macro sites and antennas. Today's macro BTS have separated the RF from the controller. It is the evolution that has made things so different. Small cells, on the other hand, are an all in one unit.

Antennas are much smarter now. They can control more and more of the tilt. They can zero in on a specific user with the signal. They could cover a wide area or narrow area. It all depends on the design needs.

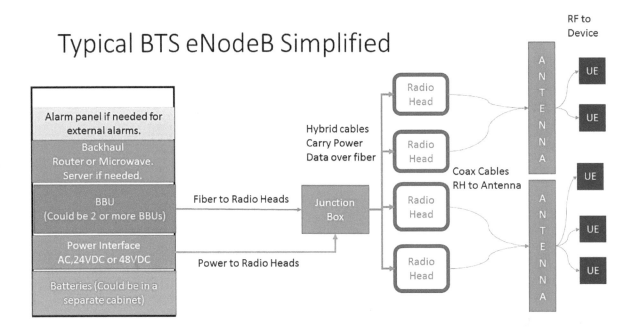

Figure 9

Wireless Deployment Planning Overview

Let's go over the basics of deployment for those of you who didn't get my last book or that did but want a quick review. First off, my last book was the **"Wireless Deployment Handbook for LTE Small Cells and DAS"** was written for the deployment teams. It was created to help the deployment teams understand the end to end process. The handoffs are very important. Something I learned over time. Why not learn from my mistakes?

Pre-deployment Planning Overview

So, when looking at the steps to deployment, what do you need to know? When you're planning the deployment, you need to understand what the plan is. This overview is to point your team in the right direction. A quick overview of what to look for when starting the initial deployment. This won't just apply to the new system but when doing expansions or add-ons. I'll bet you see most people just building out a duplicate system not thinking of the spectrum or the service they want to offer. You see, this isn't the best strategy. There are times when you need a lot more spectrum, so you need to overlay onto what you have. That's easy, but what about when you have a 600 or 700MHz spectrum, and then you have 2500MHz or 3500MHz? You can overlay them but the coverage properties are not the same, building penetration is not the same. Make sure you think it through.

Before building out the network, look at the desired result. Before you think of antennas or OEMs or equipment, what is it that matters? Your team must solve problems before making any commitments. If you already have spectrum then you know what you need to do, look at these questions and answer them for your target customer. Start with the end in mind, and the end is the customer experience.

From my last book with some updates because 5G will be open to more than just the carriers. It will be open to the enterprise users and the sports stadiums and even home theater. The medical industry and industrial repair companies could have their networks. I will give an overview below.

- Your Business Case
    - What is your business?
    - Who is the end user?
    - What technology will you use?
- Let's look at the business first.
    - First things first, budget versus coverage versus services versus throughput.
        - What is your budget?
        - What are you trying to cover?
        - Who is your designated customer?
        - Do you have a throughput goal?
        - Do you know what services or applications you want to run?
    - Partnerships, are you working with partners?
    - What is the goal of your system?
        - Whom are you servicing?
        - Are you working to cover a specific area?
        - Will you provide a specific service?
        - Are promising a designated throughput?
- System requirements:
    - Is it a Heterogeneous Network?
    - Ask these questions:
        - When do you install Small Cells?
        - Why Install CRAN?
        - Why Install DAS?
            - iDAS
            - oDAS
        - Why install Mini Macro Cells?
        - DAS or Small Cells or DAS and Small Cells?
        - Unlicensed Band and lightly licensed planning.
            - LTE-U?
            - Wi-Fi

- o CBRSD
- Carrier Aggregation
- Carrier Aggregation with Wi-Fi, LTE-U, and LTE
- Backhaul and Fronthaul
  - o Backhaul and Fronthaul Options
  - o Backhaul Planning
- o What is involved in deployment?
  - Project Management
  - RF Design
  - Site Acquisition
  - Site Survey
  - Site Design
  - Network design
  - Installation
  - Commissioning
  - Integration
  - Optimization
  - Inspections
- o Deployment challenges:
  - Installation Skills for Small Cell, DAS, and CRAN
    - Fiber Connections
    - Copper Connections
    - RF Connections
  - Mounting the Small Cell or CRAN RRH
    - Pole Mounting
    - Strand Mounting Notes
    - Stealth Mounting Notes
    - Grounding
  - Overcoming Challenges: Problems and Planning
  - RF Coverage versus Offloading
  - Permitting and Zoning Challenges
  - Backhaul and Fronthaul Challenges
    - Power
    - Mounting Assets
    - PIM Testing
    - Tiger Teams
    - Installation

Planning and budgeting for deployment.

Let's look at what goes into planning for the deployment process. When planning, the deployment plan makes sure you understand the result. Start with the end in mind. For instance, if you are going to build a network for data, who will be your target customer? Will it be "machine to machine" or will it be "fixed connectivity" or "mobile connectivity"? What will be the main source of traffic?

Start with the end in mind.

The key is to think about whom you want to serve, where they are, and what services you wish to provide. These are the keys to moving ahead. You don't need a solid commitment for any length of time, but you need to start somewhere, and you need to build a plan. That is the key to starting. Trust me, when you get moving you will be making changes but the system can handle changes. If there are limitations in the equipment you chose, push the OEM to do what you need, just be ready to pay for it.

Put some thought into whom you are going to serve.

The graphic below should show you how to tie it all together. Think of the result, then work your way backward and put all the connecting links together. In today's world, the loading and the backhaul will make a difference. You need to know what the connection to the core will require. In today's cloud environment, you may be able to lighten the load to the core by utilizing the cloud connections.

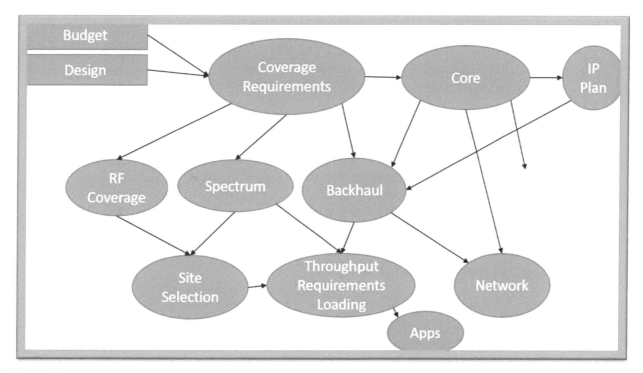

Figure 10

That's right; you no longer look at the deployment for coverage only. You should look at the very end, the apps, to help you determine the bandwidth needed to build the rest of the system. It's ass-backwards of how we used to do it. Before, coverage. Then loading. Now we look at the network system, the loading requirements, the heavy usage area, then we put it all together.

Let's look at the HetNet, the Heterogeneous Network, where we add in the macro cells, small cells, DAS networks, and anything else we can shove into the network to make it as dense as possible.

But wait, there's more! That's right; the backhaul is a critical link in the chain, so engineer it beforehand. We aren't doing just voice where we look at T1 or E1 or DS3 circuits. We need to engineer the all-IP backhaul, be it wireless or wired or fiber, to handle the expected bandwidth and make sure that it has room for plenty of growth. The networks are getting bigger, not smaller, and constantly being improved and made more efficient, but traffic always goes up, not down. Plan accordingly as realistically as possible.

What is the service?

Are you providing fixed or mobile service? Are you providing service for mobile devices or laptops or tablets or all the above? Is this data only? Is this for voice and data? Is this for people to watch videos or for machines to talk to each other. While you go through this, you decide on what you're providing then you decide on the system you need and the spectrum you need and the bandwidth you need and the devices you need. It's all need for your paying customer or your company's solution.

What is your goal for service? What is your plan for growth? Think it through and build on it.

Break it down even more.

Let's break it down to something that is even simpler to understand. Remember that this is high level and a good place to start.

1) What are we going to do? Applications and services.
    a. What service are we offering, i.e. wireless pipe over 1Gbps, a small pipe for IOT products, specific apps, low latency services, voice, etc.
    b. What apps do we want to run? Voice, telephony, video, data only, etc.
    c. Who is the target customer? I mean like utilities for meter reading, or enterprise for locations services indoors, or mobile customers walking around or automobiles. So many choices but you need to narrow it down.
2) What is the coverage?
    a. RF coverage for the customer? Territory.
    b. Indoors, outdoors?
    c. Building penetration needed?
    d. High power or low power units?
    e. Do you have spectrum? Licensed or unlicensed?
3) What do you expect the bandwidth to be?
    a. You need to know this for not only the end customer but to plan the backhaul and the core connections.

End User Application?
- Services, i.e. wireless pipe over 1Gbps, small pipe for IOT products, specific apps, low latency services, voice, etc.
- Voice, telephony, video, data only, etc.
- Target customer, user, their needs?

Coverage?
- RF coverage?
- Indoors, outdoors?
- Building penetration?
- Hi or low power?
- Spectrum? Licensed or unlicensed?

Bandwidth needed?
- To end user?
- Backhaul?
- Core?

Figure 11

Inter-Network Connectivity

With the coming of 5G, we will see more and more HetNet systems. If you are in the industry, but the great thing is that the networks will be able to talk to each other. Much like the internet now, where you just plug in, the wireless networks will start to connect to other wireless networks. The carriers will be able to connect to an independent network and hand off data. They are doing it now with Wi-Fi. However, Wi-Fi is not as friendly with LTE as we would all like to think. That is why LTE-U will make things easier to interconnect wirelessly. How exciting is that? We shall see the carrier independent small cell a multi-carrier small cell. How cool is that?

That is why you could build your network then reach out to the carriers to see if you could connect to them. The carriers made it clear that they don't want to pay for small cells or DAS unless they see a clear payback. However, I think they would entertain a partnership with a business or company that could help them serve their customers.

RAN Backhaul and Fronthaul Overview

When looking at the RAN you may not think of backhaul or fronthaul as a component, but it is a critical one. Think about it, without backhaul you have no connection to the core, and without fronthaul, you have no connection between the BBU and the radio.

RAN Backhaul Options

Microwave backhaul

- Line of Site
- Near Line of Site
- No Line of Site
- Not always practical

Fiber backhaul

- Overhead
- Underground
- To the pole
- Not always available

UE backhaul

- Available in coverage areas
- Cost effective
- Must configure donor site

Figure 12

Let's start with the backhaul. The backhaul is the connection between the BBU and the core, not a plug and play device just yet, although the small cell has that aspect of it. At the site, you need to have the components to make the connections happen.

First, let's cover what the backhaul is. Today's network will have an all-IP backhaul. What this means is that it will have an Ethernet connection to the router. The formats will all be IP based for 4G and 5G connections. LTE is an all-IP format, as 5G will be. Remember that LTE is one of the building blocks of 5G.

In the days of CDMA and GSM, what we called 3G, they had traditional telco formats like T1 and DS3. These formats worked great at that time, and they were the foundation for what telco had to offer. However, they were over copper. They had limited bandwidth whereas today, with fiber, we can get more bandwidth. When building these systems, there is a need for more and more bandwidth. While DS3 could supply up to 155Mbps of bandwidth, it took more equipment to take it from IP to DS3 format and back again, so now Ethernet connections are the standard in most carrier backhaul systems.

What do you have in the backhaul and fronthaul components? You have the router at the RAN. Chances are the router will be Ethernet in and Ethernet out. 3G systems used T1 and DS3 formats for the connection to the internet, but now all connections are pretty much IP in and out.

The standard connections could be copper, fiber, or microwave. Fiber is the most common for macro sites because they can deliver speeds greater than 100Mbps, in fact, as we go to 5G, the carriers will expect 1Gbps and up. Microwave is trying to catch up. You can find backhaul that can do 1Gbps links, but the hops are very short and LOS. You also should worry about latency, which is a real issue with fronthaul. We'll get into that later.

Then, out of the router, you will have IP access which may go to a switch to distribute the data among the components in the BTS. While the main purpose is to connect to the BBU for the backhaul, it also passes more information back to the core such as alarms and BTS status. There is also a control channel for the remote MME to manage the BTS. With the IP connection, there are so many things you can monitor and control Most OEMs already have most of this built into their alarming systems. They even look at temperature and open doors. Some carriers are running video back through the backhaul so that they can see what's going on at the site when no one is supposed to be there. However, the data to the BBU is the top priority and video is a convenience at best.

Fronthaul is a connection between the BBU and the radio. In the case of a macro site, you have a fiber run, generally in a hybrid cable, between the BBU and the radio head on or at the tower which could be a simple piece of fiber connecting the 2. The reason they call the cable a hybrid is that it will have 3 to 9 or more fiber strands running through it along with power for the radio heads. The power lines are copper lines for DC power up the tower. There can be AC power on these lines, but it would be low power, chances are DC or AC it will be 48V or less. Does it have to be big enough to carry the current to run 3, 6, or 9 radio heads on the tower? There is loss through the cable that, if the engineering is wrong then you could have problems. Radio heads need power to work.

Fronthaul at the tower is straightforward. However, in today's world, we have small cells and remote radio heads that are part of CRAN, Concentrated RAN, systems and we have radio heads that could be part of a cRAN, cloud RAN. The idea of these systems is that the controller, a BBU, will be at a remote location controlling several radio heads from that location. Generally, in CRAN they are called BBU hotels, making maintenance and control of multiple radio heads at remote locations a lot easier when the tech can go to the one location to make changes or upgrades.

So, fronthaul will have a router and most of the time it is fiber. You could also have microwave. Copper is not too common because they want dedicated connections, fiber and microwave offer that. Copper does not.

The issue with fronthaul is that the latency must be very low, there are communication timing issues between the BBU and the Radio Head and the UE that are critical. You don't want the packet to time out, so you have distance limitations with fiber and microwave. Fiber is clean and works very well. Some microwave systems have longer delays due to the conversion between the data and microwave which can be an issue when transmitting signals because if they time out, then it causes retransmissions which will cause problems in the network if there are too many. Yes, there are delays through the microwave system usually from converting from IP to RF then from RF to IP on both sides. It takes processing power and if there is a problem with the link, noise or interference, then the RF side will start data recovery and possibly retransmissions.

Let's look at the backhaul connection. You can have fiber, copper, microwave, or other connections.

Fiber connections:

The most desired connection to the core. Fiber allows a huge amount of bandwidth. Over 1Gbps of bandwidth is available with the right equipment. You have limitations, but it works well.

What options do you have for fiber?

- Dark Fiber – this is an unused dedicated fiber optic cable that to the customer's purpose. In other words, you aren't sharing it with anyone. A dedicated connection between the RAN and another site or the core or wherever you pay to have it sent. For dark fiber, the customer, you, will need to provide all the equipment to connect. You can get a huge amount of bandwidth through dark fiber, 1Gbps, maybe more. Your limitation may be your equipment. It is easy to scale dark fiber. If you run your dark fiber, it can be very expensive because you must get permits, right of ways, pay pole rents, maybe trench, and so on. It can get very expensive.

- Lit fiber – this is a shared fiber, and you connect to the carrier's equipment. The carrier could be a telco or fiber carrier or anyone who offer service. It is usually cheaper, but it is not a dedicated connection. It will still connect between 2 points, but the bandwidth may be limited because you are sharing the fiber. You may have a problem scaling up and need to coordinate with the carrier to make changes.

When is fiber used/not used?

- Macro sites that require high capacity could connect to the core or to another macro site to save on costs.

- Fronthaul for low latency and high capacity to connect the BBU to the remote radio head in a CRAN option.

- Small cell sites when heavily loaded or no other option is available.

- CRAN Hotel BBUs to connect to high capacity backhaul and to connect to remote radio heads for fronthaul creating a situation where you would have several fiber runs.

- In the case of C-RAN, Cloud RAN, it would be to connect the cell that is connected and controlled by a BBU in the cloud. New in 2016 and being tested in China and the USA.

- Fiber is not available everywhere. There are issues connecting to fiber in some areas.

- Fiber could be cost prohibitive to run to your specific site which has slowed the growth of small cell sites on remote poles. The cost of getting fiber to the pole may be more than the cost of the small cell and the installation of the small cell. That has been a problem that holds back the mass deployment of small cells.

- In some cases, you have only one fiber provider to choose from, and their costs may be probative.

Microwave Connections:

- Point to Point, (PTP) is where you have a dedicated microwave shot between to end points.

- Point to Multi Point, (PTMP) is where you have one control point connected to multiple endpoints.

- Latency varies, and it is hard to capture in a band. Why? Let's review this list:

- Distance – just like fiber, the farther the data travels, the higher the latency. In microwave, the longer the link, the higher the latency.
- Equipment – specifically the radio equipment in this case. The longer it holds on to a packet the longer the latency. The longer it takes to process the conversion from RF back to IP, the longer the latency. The longer error correction takes to complete the longer the latency.

- Spectrum, microwave can be in many spectrums that serve many purposes. High-level explanations for the US market but they could apply to the world. These are the most common. Remember that the distance and dish size and engineering will affect throughput and latency.
  - 6GHz range – general for long range shots. Point to Point LOS (Line of Site) microwave using larger dishes for longer shots. Licensed. Used early on, but the limitations in bandwidth and the large dish size have made them less attractive to modern sites. The dishes are generally over 6 feet and over. However, the FCC will allow 3-foot dishes in some situations. The limitations are the spectrum, licensing, and potential interference. The FCC did allow larger channels, but the current licenses in the US make it hard to get larger channels. Antenna size is an issue, but because the propagation of 6GHz is great, meaning it can travel far, it makes it hard to license without causing someone else problems. It was great with voice channels when they could travel great distances. Public safety in rural areas relies heavily on this because many of their sites are spread out.
  - 11GHz range – generally used for midrange shots. Point to point LOS microwave using mid-size dishes, around 4 foot or so, but the FCC will allow 2-foot dishes. Licensed. Used extensively I the past and is a good midrange solution. The FCC was going to allow smaller dishes, but this band usage is high and very dense in the USA. The throughput is just over 200Mbps if properly engineered.
  - 18GHz range – generally used for short to midrange shots. Point to point LOS microwave using 1 to 4-foot dishes. Licensed. These are an attractive solution with high bandwidth. Do the engineering because these links are heavily affected by weather, specifically, rain. Bandwidth through these links could be 100Mbps up to just over 300Mbps
  - 23GHzrange – generally for very short hops. Licensed. Point to point LOS microwave using smaller dishes, around 1 to 4 foot. High throughput, 100Mbps and up. Very prone to rain degradation. Very easy to license in the USA.
  - 24GHz range – generally used for short hops. Point to point LOS microwave using 1-foot dishes could go down to 8 inches. Not licensed, very easy to license, having a throughput of 100Mbps. Some companies can get this band to over 700Mbps with proper engineering, but rain is a factor when it comes to engineering these links. Very limited on distance. Interference is usually low because of the propagation properties of this spectrum. This spectrum is good for short hops.
  - 2.4GHz and 5.8GHz range – the ISM band used for short hops, (although I have seen companies connect 15 to 20-mile links). Could be PTP or PTMP. Could be LOS or Near

LOS or in some cases non-LOS. Not licensed. This sub 6GHz license free spectrum is a popular choice among non-carriers because the spectrum is free and the hardware is cost effective using smaller dishes (or panels) which are easy to install and setup. No license makes it easy to deploy anywhere, and the low-cost equipment makes it affordable to deploy anywhere. A short hop solution but there are claims that are using the right size dishes that it can be a long-haul solution. The downside is that it's prone to interference because anyone can put them up or any Wi-Fi hotspot may affect it. They are easy to deploy. Throughput varies on the engineering but generally, 10Mbps to 150Mbps. I have seen more throughput, but it takes the right design and engineering to get it.

- E-band 71-76GHz and 81 – 86GHz range – generally for very short distances, prone to weather issues. Dishes are very small, under 2 foot. Point to point hops. Licensed links, but light licensed, so getting the license is very easy in the US and Europe. These are a popular choice for short hops that could need up to 1Gbps of throughput. Very high throughput looks like a fiber connection.

- 60GHz – generally for very short hops. Point to point, but there is talk of a multi-point product coming out. Dishes are 6 inches to 2 foot. Throughput is very high, over 1Gbps.

When is Microwave used/not used?

- Microwave is a cost-effective alternative to fiber, but can only be used in specific cases. Your paying for the hardware, so CapEx is higher. The OpEx is lower because the only reoccurring cost is license renewal and tower rent if you're paying it, and maintenance.

- Microwave works for macro and small cells for backhaul or fronthaul.

- Microwave does have its drawbacks because it is a limited solution, although a very cost effective one if you're looking at OpEx.

So, when looking at fronthaul or backhaul you have:

- Router.

- A connection from point a to point b, fiber, microwave, or copper.

- Switch (if needed).

What is LTE UE backhaul?

It is backhaul that uses the carrier's spectrum, just like the UE, User Equipment, your smartphone. If you have ever used your smartphone as a Wi-Fi hotspot, then you know the concept, using the carrier's backhaul to create a new hotspot. Now imagine taking your usage and multiply by hundreds or thousands of megabits. The UE backhaul device in something that will use the carrier's LTE spectrum for backhaul. This is something that is commonly used for internet access when there is no Wi-Fi available. The carriers all sell these units and many of today's smartphones do something similar. However, they just use the standard signal. Using it for a tiny hotspot and for an eNodeB are 2 different things.

The 5G Deployment Plan Handbook

Let's talk hotspot. Many vendors provide equipment that a user can add coverage quickly and easily. It is a quick Wi-Fi connection to the internet using the carrier's LTE to connect to the internet. Everyone has Wi-Fi, and there are devices that create an instant hotspot. Verizon has the Mi-Fi, or you can use your smartphone as a hotspot. Every carrier has a wireless modem that you will provide a Wi-Fi hotspot. I think anyone who is reading this knows about the hotspots. I thought it would be a good example to get started.

What is a cell extender? There is a practice where many carriers will use a cell extender that will have a UE relay backhaul to extend the signal. This is also like a smartphone hotspot or a Mi-Fi unit because it was just to help a few customers but extends the carrier's signal instead of Wi-Fi. This is a type of repeater to extend the macro's signal, a cell extender. This is a way for the carrier to extend the coverage just a little bit farther. It's a way to provide coverage someplace quickly and easily. These were common in 2G, 3G, and now LTE. It is a simple and quick way to install a repeater to extend carrier coverage down an ally. In the old days of DAS, this is what they did. They would take the signal where it was strong or use an antenna and amplifier to increase the strength to get it into a dead spot. People paid a lot of money for these systems.

It's not a simple cell extender, and let me tell you why. Now you are talking about putting the small cell in an area where there is a loading issue. This goes beyond coverage. The data and spectrum usage could go through the roof! If you set it up like a cell extender with backhaul to the macro site, then guess what! You will see an overloaded macro sector! The macro not only has to deal with all its users but all the small cell or Mini macro users too. This sucks up all the spectrum and bandwidth for that sector. What can be done? Read on!

I am bringing this up because now there is talk about using the UE backhaul for small cells, mini-macros, and macro cell sites. It's making a more powerful cell extender. It sounds like a great idea on the surface. This is a cheap, quick and easy backhaul. However, what is the drawback? It's not as easy as you think, the carrier needs to set up the donor site properly. I mentioned it earlier, and it is not something you just throw out there to feed a cell site. It draws a ton of data. It sounds like a great idea on the surface. It looks like a cheap, quick and easy backhaul.

The donor site needs to break the bottleneck. You need to dedicate spectrum in the macro eNodeB that will be feeding the UE backhaul. This will alleviate the spectrum usage for the regular users on the macro sector. We don't want them to get knocked off if the small cell US backhaul overloads the macro. This will make it so that the users on the macro don't get shut knocked off if the small cell pulls the entire spectrum of its users. This will allow the small cell UE backhaul to have a dedicated pipe. It needs to have dedicated spectrum for this purpose. Then the small cell will know how much backhaul spectrum it has available. By the way, not an easy change, changes in the eNodeB and possibly the core need to be considered as well as neighboring sites. This "dedicated backhaul spectrum" needs to be set aside for this sector and others too. It takes some planning and changes.

You could still have the data bottleneck at the macro's backhaul. That's another issue that needs planning.

So now you dedicated part of the band to the UE backhaul, which seems OK. Remember that the carrier paid a lot of money for that spectrum and now they are choosing to use it for backhaul. The pipe is limited based on coverage and availability. It is a quick and easy to add UE backhaul, but is this the best use of the spectrum? Will you lose something in this backhaul? Yes, you have delay issues, timing issues, and neighbor issues. All of this is a problem when building a site for any type of real loading. Go to the links below to learn more.

However, what's the real issue? Is it all the problems I mentioned above? They are all technical issues that good engineers will resolve. This appears to be a cheap and quick solution. But that's not the real issue, is it? The carriers paid a crap ton of money for spectrum. Is backhaul a smart way to use this resource? Is that billion-dollar investment there to save some CapEx for the company? I thought it was for the customers! Backhaul could have been something in the unlicensed band for a lot less money. It could be a fiber link for more money. Is this an easy out or will it cause problems down the road because the spectrum is only going to get more and more valuable? Do investors want to see that spectrum used this way? I don't see the auctions being a cheap alternative to providing backhaul.

So just because it looks cheap and easy doesn't mean it's a good move strategically. Don't get me wrong; the UE relays, the repeaters serve an important purpose for coverage and filling holes, I am just saying be strategic and think it through. For more information hit the links below to learn about these solutions.

If delays were lower, this would be a great technology for fronthaul, now that would be something!

Resources:

- https://www.nttdocomo.co.jp/english/binary/pdf/corporate/technology/rd/technical_journal/bn/vol12_2/vol12_2_029en.pdf
- http://lteworld.org/blog/introduction-relay-nodes-lte-advanced
- http://wireless.skku.edu/english/UserFiles/File/1569472705.pdf
- http://www.ericsson.com/res/docs/2013/lte-in-band-relay-prototype-and-field-measurement.pdf
- http://www.interdigital.com/research_papers/2012_01_13_system_architecture_for_a_cellular_network_with_ue_relays_for_capacity_and_coverage_enhancement
- http://www.airspan.com/products/airvelocity-2/

RAN Site Components

Let's cover the hardware that you would see at a typical site. I am going to break it down by the type of site. You see the basics that have always been there. These are all parts of the BTS. Let's look at what you have at most every site.

- The backhaul/fronthaul components:
    - Router, could have a switch or other components.
    - Fiber, copper, microwave, UE backhaul, or satellite connections.
- The BBU.
- The radio.

- The antennas and jumpers.
- Power source from AC or DC.
- The mounting hardware and structure.
- Options
    - The server could be for fog computing, testing, or on-site services.
    - Battery backup.
    - Backup backhaul
    - Alternative power like a fuel cell, generator, solar, etc.
    - Security cameras, site access controls, external alarms, etc.
    - Shelter or outdoor cabinet.

Things are changing at the site. Now we need to look at what skills are needed. Many of you already know you need fiber skills at every site.

BBU.

The BBU will be on the ground at a macro site. I put it like that because there are several variations. The small cell could have the BBU and radio all in one box. The cRAN and CRAN could have the BBU at one location and the radio at another.

So, what is the BBU, the baseband unit? The brains of the radio. The baseband unit will interface between the radio and the core. The BBU takes all that data and processes it, converts it to a signal that the radio head can use so that it talks to the UE. The BBU talks to several components in the core like the MME, (Mobility Management Entity). It needs to know who can connect and who should not. It needs to take all the data from the UE and send it out to the core the out to the internet. It may have to route it to another cell site to handoff. It may have to process voice and data.

Radio.

The Radio is generally in a radio head. Fiber jumpers connect the BBU to the Radio Head.

Antennas and Jumpers.

Antennas are quite complex today, and they also have many ports to support many bands and possibly the uplink and downlink.

The Mounting Structure and Hardware.

The hardware includes cabinets, mounting hardware, miscellaneous items. You may even need to have stealthing for your site. Also, with the site survey you can work out the details

Battery backup.

Having battery backup may be required at Macro sites, but for the small cells, they are a luxury. The government in the US has regulation on battery backup for carriers. They do require batteries or some backup power so that your smartphone works when the power goes out.

Testing at the site for more than the equipment!

Make sure to test the cables at the site to avoid a return trip. The cables need to be tested, a step too often skipped because the tower crew may not have the expertise and the carrier just don't want to pay for it or take the time out of the schedule to verify that the cables and connections are good.

RF Cables – the coaxial cable and antennas should be tested for passive intermodulation interference, referred to as PIM. A phenomenon that happens at the connectors in RF spectrum. There is also regular intermodulation which is interference from other bands when they mix. As the carriers move into broadcast frequencies, they will see more and more interference from other carriers mixing with broadcast bands. Most PIM problems are due to a faulty RF connected or an improperly installed connector. Connectors can be repaired quickly or replaced during the installation. It is much more money to go back out, climb the tower, and replace the cable later.

 Fiber Cable - Fiber testing is when they test the fiber lines between the BBU and the RH. These cables are fine if they are factory mad, don't have any sharp bends, and are connectorized properly. With the growth of fiber on the tower, the tower crews had to learn new skills like terminating fiber cables. Fiber cables need to be clean to work, not an easy task on a tower. Testing is imperative for the workers. Fiber cables may have bit errors when they try to pass data, which could be the result of a speck of dust on the connector. Like the RF cables, this problem can be addressed quickly and efficiently during the installation instead of having problems down the road where the tower crew must go back out and do it later.

The 4G deployment plan

Planning is key for the deployment to be a success. You probably think that it's up to the carrier to plan everything out. You think that all you must do is your part to help lay in its success. You would be wrong. If you want to get more work, then you need to play a part in the planning. For instance, do you know what your entrance criteria are? Do you know what your hand off and close out criteria will be? If you define this up front, you are showing that you know what needs to done and that it's a team effort. It is something that more people won't worry about until they are on the job.

The other advantage to each team looking at this when they bid the job is that they know what it takes to be profitable and they know that most carriers or general contractors, GCs, will be looking to get as much work out of you that they can. If you define that up front in your scope of work, SOW, then you can bid the job properly and profitably. It is something you need to plan. So, let's break it down into each part of the deployment.

Types of Cell Sites

There are many different types of sites. I discussed them in my last book The Wireless Deployment Hand Book: The Small Cells, CRAN, and DAS edition. However, for the purpose of this book, I should cover them again. The 5G sites will include a similar architecture, so I am going to specify what will be likely for each of them in the statements.

Macro site - can be a BBU, router, Radio Head, and all the parts in between. It is a full-blown site that will have a large backhaul and could serve hundreds of customers, perhaps over a thousand. The sites are at least three sectors and usually have several carriers at the site. They have larger antennas and higher power radio heads. They usually have a full blown BBU and cover enough spectrum to handle many customers. These sites are usually on towers or rooftops. The higher power radio head means that you don't want it too close to the general population, usually a hundred feet up or so to be safe in the USA. These sites have large backhaul, usually fiber, to the core. They must carry a lot of traffic normally, but these sites are also built in rural areas to cover a large area with higher power radio heads.

Small Cell – Small cell are low power cell sites, usually 10 watts or less, that have the BBU and radio head all inclusive. They usually are one unit, but people often call CRAN system small cells, I will explain below. These are used to hit fewer customers than a macro site, anywhere from 30 to 300, in a concentrated area. They could have an Omni antenna, but generally, they have a directional antenna and concentrate on a specific small are where people gather. They could be indoors, which is usually less that a 5-watt radio head, or outdoors, usually 10 watts per radio head or less.

CRAN – The Centralized RAN is a term used when the carrier locates a BBU in one area, maybe a closet in a build where fiber backhaul is available, and the radio heads are located with throughout a building or a town. This BBU is usually called a BBU hotel because it controls several remote radio heads. The advantage of this system is that it's easier to manage. That means that you can control the spectrum, loading, and neighbor list on each radio head from one BBU. The BBU is easy to maintain since it is in a central location making maintenance easier because you can push firmware to all the remote radio heads. It is an efficient way to manage sites. Now, it is not that easy because there are limitations or the fronthaul that connects the BBU to the remote radio head. The fronthaul must be very low latency and have plenty of bandwidth. Dark fiber is very popular for this application, but microwave is an alternative. While not all microwave vendors have low latency, that becomes an issue. Fiber, as you know, is not cheap, and dark fiber between 2 specific remote locations may cost even more to engineer and monthly. All considerations when going this route. In the USA Verizon, has been very happy with this system and they continue to move ahead with the CRAN approach.

cRAN – OK, I know this looks the same as CRAN, but the small cell signifies Cloud RAN. Here the BBU could be controlling the radio head from the cloud, not from a normal BBU hotel. Now, end of 2016 beginning of 2017, being tested in China so that they don't need to put any equipment on the ground except maybe a router.

Mini Macro – The mini macro is a bigger small cell, could do 20 to 40 watts. The reason for the mini macro is because it is easy to install and integrate than a macro, but it has more power and can handle more users than a small cell. So, the mini macro is a good in between solution. It will have limitations, but when a small cell isn't enough, but the carrier doesn't want to spend the money on a macro site, then they can install the mini macro.

DAS – Distributed Antenna Systems are another solution for indoor or outdoor coverage. Often cRAN or external radio heads are used for the LTE solution with antennas distributed all over for coverage where needed. Unlike the analog DAS systems of past, this can't be a simple repeater. It's a digital system that will have a BBU or some way to connect to the core and then it will work as a cell site.

The BTS Installation.

A ground crew will install the BTS because it is very heavy but delicate. Installation is more than the cabinet. Then the BBU needs to be commissioned and powered up. When I say heavy, it can range from 100 lbs. to over 300 lbs. these units are heavy and hard to move and delicate, they can easily be damaged. So, the installation crew must be very careful when working with this equipment.

You could be installing it in a shelter, which was the preferred method used up until the 2000s when the tower owners started charging for real estate you used on the tower and the ground. They started charging for every inch, which makes them more money. However, this has forced most carriers to move away from shelters by using outdoor cabinets. It makes sense, but the equipment had to be made to work in an outdoor environment. Also, when the technician or engineer need to work on the equipment, they are outside unless they set up a tent. It has made working on the equipment in harsh weather unpleasant, to say the least. I remember doing it, and it was a pain when you had to get test equipment out and work in the snow, rain, cold, or high winds. It's not fun, let's just say that.

When doing the installation, the carrier will have a MOP, method of process, on how they would like the equipment installed and mounted. Whether it's inside or outside, you should know exactly how to install it and how to anchor it. Indoors it's easy to anchor it to the floor, but outside you will have a concrete pad. In that pad, you need to anchor the equipment. You also need to worry about sun, water, and snow. The vents must be high and clear. Think ahead, do the planning to make sure that you won't mount it in a way where water can get in or snow could block the vents.

One other thing to think about when mounting outdoors is the sun and heat. Most cabinets have a sun shield to help keep the unit cool. It keeps the direct sun off the cabinet so that you control the heat without a cooling unit. It could be something simple that makes a big difference in the trying to regulate temperature in the cabinet.

Think about the surroundings. The equipment should be in a compound. What I mean by that is it should be in a fenced in area where it is secure, or at least as secure as a remote site can be. Be aware of where it is mounted and how protected it is.

Power is an issue. When mounting a new cabinet for the first-time power must be thought about ahead of time. Chances are good you will have AC power coming into the cabinet from a power panel into the cabinet, probably a sub panel in the cabinet. It may go directly to the power supply where you have your breaker on a DC breaker panel. All the same, everything inside the cabinet should be wired and ready and completed by you or the commissioning engineer.

The Radio Head Installation

Many people think if you put something in and hang some antennas then you're done. For those of us in the industry, we know that there have been big changes in the past ten years. Even with 3G, the radio heads, (RH) were often on the tower. You can put them on the ground, and sometimes a carrier does. More on that to come. For the most part, the RH is on the tower mounted very close to the antenna. At least for Macro sites, they are this way. I will have a breakdown, but we should start with the Macro sites first.

Antenna Notes

One thing that has been a hot issue is the antenna technology. For 4G antennas have taken a great leap for many reasons. They not only have multiple polarities but they also have multiple bands and technologies. It is becoming critical to have the antennas handle more for several reasons, listed below.

- Carrier aggregation will start to pull several channel and spectrums together, to do that they all need to reach the UE device at the same time in relatively the same strength. Antenna design will help make this happen.

- Tower space is becoming an expensive part of the puzzle. The more you can do with one radio head and one antenna, the more you can pack into a small space for the same amount of rent money.

- The weight of the antenna also is an issue because the towers can only hold so much weight. Companies like Sprint and AT&T are using older towers that can't support the weight. So, what they do is mount the equipment, radio heads, on the ground and run coax up to the antennas. Some towers can't hold the weight, so it's cheaper to put it on the ground.

- Size also matters for the tower, the smaller, the better.

- Small cell antennas need to be very smart, what that means is that they will focus in on specific devices to allow more bandwidth and better connection. Antenna manufacturers have found a way to focus the energy on specific devices, cool. Massive MIMO is a function that requires smarter antennas.

LTE MIMO Deployment Notes

I found a video that is interesting if you are into massive MIMO. It's proposed for outdoor work. I think this is interesting because Professor Dr. Wolfgang Utschick talks about how MIMO works. He gives a long and detail explanation (snooze). I listened to it because I find it interesting. (Welcome to my Saturday mornings, seriously).

The video, https://youtu.be/zhncADqR9rg, goes into detail about the complexities of how the MIMO works down to the signal level. A smart guy giving a boring delivery so that I will break it down for you, if you listen to my podcast, you may find it more interesting than the video. Let me tell you my version, lots of antennas = better signal propagation both ways, with better noise rejection and more throughput. Multiple signals are going in and out simultaneously, allowing the device and BTS to work better by cleaning up the noise and errors so that the customer can get some kick ass bandwidth. That is the name of the game. Then he talks about the multi-user MIMO works using the same signal. Then he sums it up by going over the beam forming properties of the antennas. What does this mean to you? Well deployment teams, it means that the RF designers will be working with multiple antenna systems. It means that the site designers and the site acquisition teams now must work with MIMO antenna systems either on a building or an antenna or small cell or DAS. Yes, they will be deploying these for DAS. Don't think it's something new, look at what Wi-Fi has been doing for a few years, and they are big into MIMO. Now they want LTE to do more do more with one antenna that years ago, couldn't happen with two antennas, up to 8 or even 24. It's balancing practical installations to what your device, (smartphone) can support. They must work together after all.

Let's not forget the installation teams will be dealing with larger or heavier antennas and more cables on the tower. That's right, bigger and heavier. What about the remote radio units, they were just starting to get smaller, and now they will be bigger, or they will add more. You will find out soon with 4T4R and 8T8R.

Then there is the optimization where the drive teams will need to get new devices to test the coverage and throughput. So, this will add complexity not only to the system but the testing as well. Just like with carrier aggregation, the MIMO upgrades will make things more complicated. What will the carriers say? Well, they are already deploying 4T4R, four transmit, and 4 receive MIMO, and some are doing 8T8R. They are working their way to 16T16R. How cool is that? I believe they will push to do more if it is cost-effective. Some carriers saw this as a ploy to the OEMs and antenna companies to sell more equipment until they saw the payback. Yes, the payback of efficiency and bandwidth. They are going to do all that they can to improve the pipe, like this and carrier aggregation. All ways to get the biggest bang out of the bandwidth they have.

To get the most out of this, they need to shrink coverage areas as well. So, in doing this, they may not need to maximize MIMO. I think to find a balance between the cost for MIMO, and the cost to deploy a site will maximize the investment. They want a reasonable coverage area based on loading. In the old days, it was based on population, but now in the world of data, it's a balance of population and usage. Now the carrier's system is becoming more and more of a pipe. They know they can't do it all, but they can provide quality coverage to the mobile masses. They must do this within a budget.

The goal is to make sure the user has a great quality of experience, (QoE), for the right budget. Of course, it could be better but at what cost? The equipment and the services start to run up the CapEx and to maintain something like that may run up the OpEx although I am not sure how. The only thing I see is backhaul will be bigger, equipment maintenance, and maybe tower rental. There may be more that I am missing.

How does CapEx go up? Let me tell you the obvious. The hardware goes up. The antennas cost more. The radio heads cost more. Chances are the BBU, and the hybriflex cable will cost more. It all adds up. They to install it, extra weight, extra testing, extra optimization, and all the little things all add up. It ain't free! All those nickels and dimes add up to hundreds or thousands a site.

So, when looking at the new LTE systems, now you see the complexity that is in a simple design. You also see that budgets play a part. Not every carrier can throw money at these issues, but they will do what they can to serve the user and to have bragging rights. Going to LTE gives them bragging rights, doing VoLTE also really helps.

There is a long-term goal as well. If carriers can get the LTE system up and running, then they can start to decommission 2G, and 3G systems are saving on maintenance and service two systems as well as freeing up that bandwidth for 4G. Get the old systems out, maintain the current system, and save money while increasing the QoE for the user. It all makes sense to me.

What about 5G? Well, from what I have been reading is that the 5G will be an extension of what they have now. I know that the carriers do not want to start swapping out gear so soon, especially in 2020 if they don't have to. They want to just an add-on to what they have, or they want to do it all through software upgrades. Why spend the massive amounts of money if they don't have to? After all, we went from 3G to 4G in a very short time. Why not use MIMO and other ways to improve the system? It all makes send to me to have the hardware ready for software updates. Let's decommission the 3G system before we replace the 4G equipment.

I am hoping that 5G will change the IOT, meaning machine to machine where we can get real-time readings for our power meters, gas meters, and water meters. I know that they have this in some parts of the country but not where I live. Hell, they don't even read the power meter every month, so if I have something in my house, that is sucking down power I don't know about it for two months! Just venting here but I see great things happening soon. I know the utility companies are waiting for federal grants to move ahead, but come on! I can see my bank statements and credit card bills in real-time. Let's get started on making the meter reading happen in real-time.

From 4G to 5G.

What are the driving factors to take you from 4G to 5G? I am so glad you asked! We all think that it's about bandwidth, and it is. However, wireless networks traditionally have limited bandwidth and other issues. It took decades, but now the bandwidth issues are starting to be resolved with technology and with more spectrum making things better from that aspect, and although there are still limitations.

For instance, **latency** in the network is response time for a packet to travel round trip from its origination point to a destination then back again. A good definition is here, http://whatis.techtarget.com/definition/latency if you want more information on types of latency. For this book, I will cover simple network latency.

The 4G systems will need to carry more bandwidth than ever before because they will be the foundation for 5G on our way to a new generation in wireless.

Remember that carriers will retire 3G and use the spectrum for 4G and 5G. 3G is not moving forward. It was a network that will be sunset and tossed aside. LTE was supposed to be different just like the name implies, Long Term Evolution, where it will be the foundation, the support, for future generations. So, when you hear of 4. 5G and 4.9G on the way to 5G, it all has LTE to thank for making it move much quicker than previous evolutions.

The 5G Pyramid

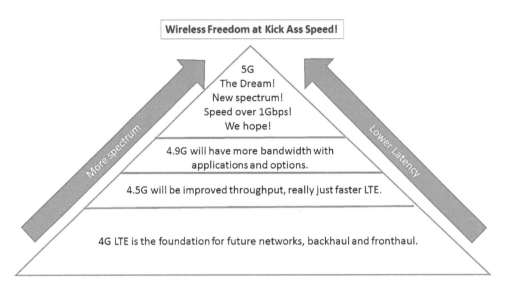

Figure 13

We all want to reach the maximum bandwidth with the lowest latency, or at least that is the goal today. It could all change in 5 years. Remember that just ten years ago, we all wanted to hear a pin drop over the cell phone. Now we want to run hundreds of apps in a few minutes.

The 4G and 5G HetNet

Remember all the talk about the HetNet, the heterogeneous network? I show it below is the HetNet using all licensed components, However, for the HetNet you can have unlicensed connections like Wi-Fi and LTE-U. It is the combination of many different types of BTS, from macro to mini to small cell to a hotspot.

The way that LTE is going to migrate into 5G is by using more and more different types of RF units and technologies. Many RF bands will be working together to give the end user a great quality of experience. The end user will not know that all of this is going on if it all works well. They will notice problems. The Quality of Service, QOS, will matter to some customer depending on what application they are running.

When the customer starts seeing problems, it is a Quality of Experience, QOE, issue. You want every user to have a great QOE because they will switch service if they don't get it. However, it does matter what they are doing with their device. Most users are very forgiving with data problems. They expect to have some downloading issues.

I bring this up here because switching of services will great affect the user experience. In an unlicensed spectrum, there is the potential for interference which will cause problems with the QOE because it won't have a clear signal. For example, if you use your Wi-Fi at home, chances are good it works well in your house, right? How about when you use it in a public area where there are many hotspots? Then you see a serious degradation in service. They interfere with each other while the UE equipment will jump across hotspots.

So, the HetNet will be a bigger part of the future networks, but the carriers will need to smart about the handoff. They shouldn't hand it off outside of the licensed network unless they have no choice. Today, 2016, there is a problem with overloading, so the carrier welcomes the unlicensed network. They want to see LTE-U because of the format and the improved efficiencies. I'll talk more about this later.

The HetNet

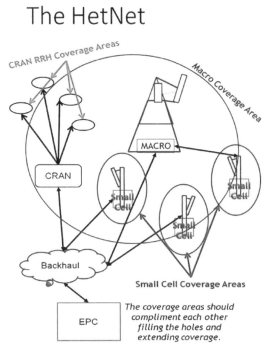

Figure 14

- The Het Net doesn't just include macro and small cells, but it could also include Wi-Fi. It is all the devices. Just make sure you have them all in one system working together, like an orchestra playing all the different instruments into one beautiful song. That's right, they all work together. Take a look at the picture below and you will get an idea of the Het Net coverage.

- Per Google: *"A **heterogeneous network** is a **network** connecting computers and other devices with different operating systems and/or protocols. For example, local area **networks** (LANs) that connect Microsoft Windows and Linux based personal computers with Apple Macintosh computers are **heterogeneous**."*

What will 5G networks look like?

Here is what I see, the carriers are stubborn to keep the LTE systems in place, they don't want to build another system for another ten years. Do expect them to improve LTE, which does stand for Long Term Evolution, for another ten years or so. Luckily, they found ways to improve the data rates through aggregation and the FCC is releasing more spectrum. It's great for the carriers who dominate spectrum ownership. Why wouldn't they when it costs BILLIONs just to attain spectrum. I don't see many small guys doing much with high bandwidth.

We have all been hearing about 5G but let's look at what's going on. It is not going to be just another new transport format for the carriers. I would look at it as an "add-on" of smaller networks giving us a new look at the HetNet. HetNet is a term we just don't hear as often as we used too, but it is the foundation for the 5G.

The FCC is releasing very high spectrum for 5G, and Verizon is testing this. How will companies attain this spectrum? What will it cost? Not an issue right now but Verizon did set a standard which they would like to see approved. I give them a lot of credit for doing real testing and setting real standards for something that may happen before 2020, the original date we thought we would see it.

Here is where IOT and 5G should merge. They should be working together as a system. However, the IOT may not be the heavy bandwidth hitter we thought it would be. Instead, it will be like it always was for Machine to Machine, M2M. Think of SCADA systems. SCADA, Supervisory Data Control, and Data Acquisition. Utilities have been using them for years. They called them SCADA. Now with the new low-bandwidth networks coming out, they may be able to start using new services if they are secure enough. They seem to like the 3.5GHz WiMAX systems because they owned them, but WiMAX didn't work as well as they had hoped. So not they are looking for a new way to handle these low bandwidth needs. Technically it's IOT, but wouldn't it be 5G? Could IOT be a part of 5G? Hell yeah!

5G will be faster. Built out in smaller networks. The 5G network will utilize much of the 4G network, but mostly to connect. LTE will continue to get faster and grow. Most of what you hear about is the high bandwidth usage, like virtual reality and the cars that will drive for us. Part of 5G, but the network will be reliable. Latency must be very low so that the cloud will matter. How? NFV, network function virtualization, and SDN, Software-Defined Networking, come into play. The process needs to be taken to the edge, as close to the user as possible to lower the latency.

I see 5G and IoT tied together. The network will start touching everything. Verizon is big into setting standards for 5G, see more here, https://youtu.be/XFjmrzw-9EM to see what they're up too.

So, the 5G network will bring the network out to the end user. With the higher spectrum, it will be a very small network. The wireless will not be the only factor; the servers will need to be as close to the edge as possible

- The 5G will have a fast wireless format.
- It may be a server close to the edge. It will need to keep latency as low as possible.
- The networking and routing will need to be very nimble and change very quickly.
- Low latency will be the key.
- High throughput will matter for some apps.
- Low throughput will need low latency.

Here is some food for thought. I put these very simple drawings together hoping it may show you how the network could look.

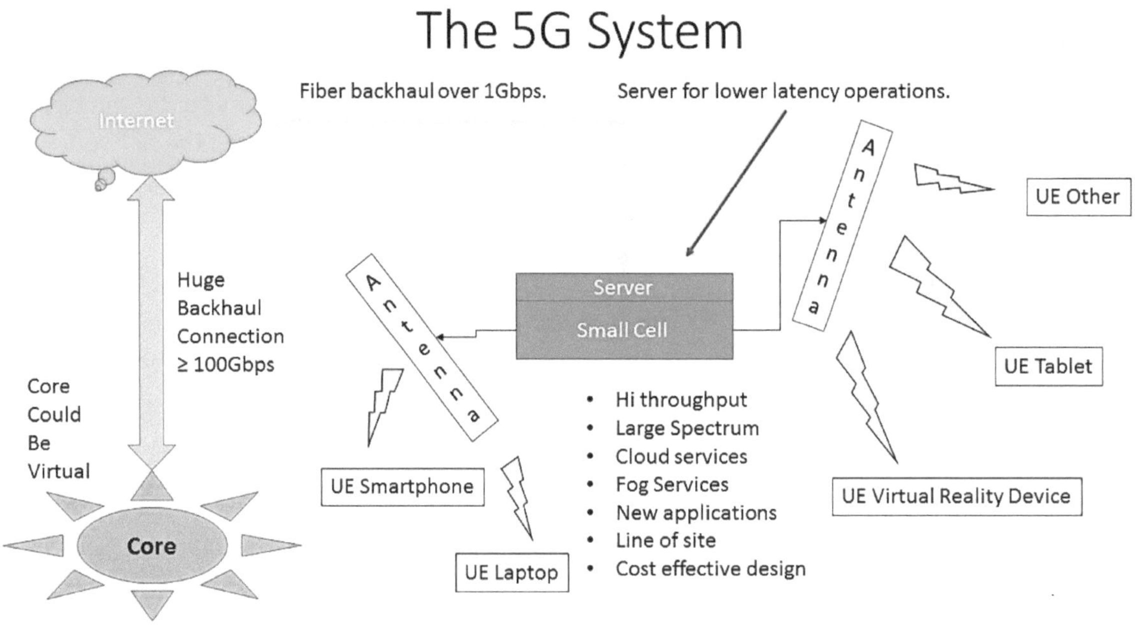

The 5G System

Figure 15

The 5G System Breakdown

System Outline

The system will consist of the following:

- Small Cells
- Backhaul
- Routers
- Servers
- Antennas
- User Equipment, (UE), devices

The breakdown of the 5G system will need to be thought out by the OEM and the carriers. They already have ideas of the 5G format, which appears to be a type of LTE. While they work out the transmission scheme, the system will be similar to what we see now.

Computing power at the small cell will be critical for the 5G system to work the way that the carriers plan to have it work for high spectrum uses. They will need to have computing power there that was not needed before. It's for the end user, not for the BTS to operate properly. The BTS should have all the power it needs to work remotely. The OEMs will enhance what they already have deployed to make it more efficient and faster.

Antenna technology will be better than before. The antennas will be made up of very small arrays. They will need to be able to focus the signal to each UE for the most efficient spectral efficiency possible. Since they will be in the millimeter wave bands, they will need to figure out how to keep connected. The antenna will be crucial because it may have to have multiple bands of spectrum in a very small area. It may be required to do multiple polarities at one time so that the device will remain connected with a strong signal no matter how it's positioned. The signal will come into play when the unit is moving, or it is near the human body where signals are causing distortion.

An issue at the other bands as well when the carriers started using 1.9GHz, and 2.1GHz ranges it was an issue was overcome with antenna technology as well as the improved error correction of the BTS and UE. It's amazing how far we have come in such a short time.

MIMO will be in use here as well. MIMO will be crucial for all of this to work, Multiple In, Multiple Out, as in multiple RF signals processed in the band being transmitted at once, 2 or 4 or 6 or 8. The receive signal does the same thing which is a key to increased throughput.

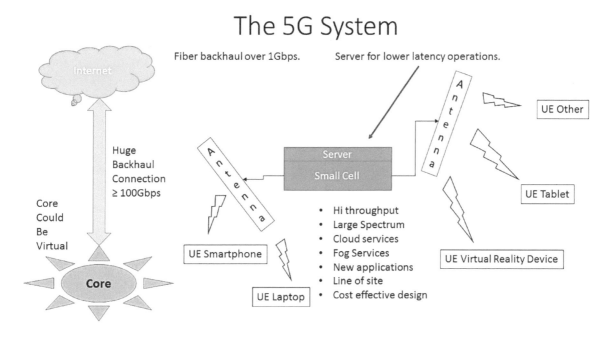

Figure 16

What is the 5G System Plan?

I have heard a lot about 5G and the roll out and how small cells will boom with 5G. Let's look at the facts, if 5G is in the millimeter wave spectrum, then they will be very small and very line of site networks.

The 5G Network

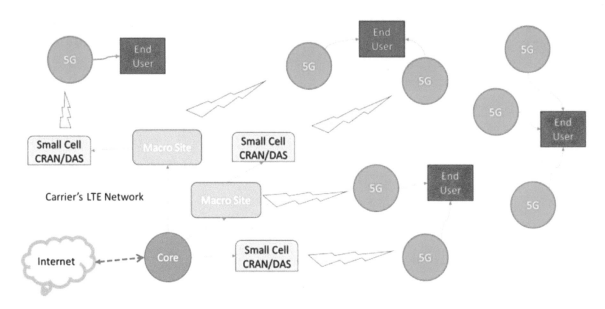

Figure 17

How is this going to work? Are the carriers going to rely on 28GHz, (here in the USA), to cover much more that a building or a street? They are complaining about putting up small cells and CRAN because of the fiber and rental fees, and the slow payback! It's catching on now, and there is a better business plan, but will it be enough at 28GHz?

What I don't' get is that they are all doing amazing research and getting awesome speeds, but to me, someone who has done point to point (PTP) microwave, it is not a surprise. If you only have 2 or 3 users, of course, you can get kick ass speeds in a high spectrum with lots of bandwidth that's not shared. I'm the only one on my Wi-Fi hot spot which works great. With a lot of users, the cable modem is the bottleneck, not the Wi-Fi, however, when I am in an airport using the free Wi-Fi with 100 other people, I have no idea what is the bottleneck. All I know is that I can't connect, I always suspect it is the Wi-Fi. It happens all the time.

So, let's look at the business model. Do you see AT&T and Verizon building a network for ten people? Let me put it to you this way when you want to put a small cell in your building for ten users, will AT&T or Verizon run out and give you one? I will tell you from experience, the answer is no, but they will let you buy one, but it's very small. Now, if I want to put something in for 100 users, will they do it. Again, the answer is no, and now because you want a bigger small cell or 2 of them, they won't even talk to you because they are afraid of how it will affect the network. I get it, but now let's think about how they will deploy 5G, will is really for the mobile user? It doesn't make sense to me.

Now, let's look at fixed wireless, here is an awesome application. One, you need LOS, (line of site), but you have the spectrum and the ability to put a small cell near people's homes. I see it as a great asset because finally, would give some wireless competition to the Cable modem.

The 5G Deployment Plan Handbook

For fixed wireless, it looks like an awesome solution, but they should make it work for the mobile device somehow.

The reason I am not so positive is that if you look at path loss, you may see something negative. At 200 meters, there is a lot of free space loss here.

Table 1

Frequency	Distance (Meters)	Tx Output (dB)	Rx Gain (dB)	Free Space Loss (dB)
600MHz	200	40	3	31.02
700MHz	200	40	3	32.68
1.9GHz	200	40	3	41.04
2.4GHz	200	40	3	43.06
2.5GHz	200	40	3	43.42
5.8GHz	200	40	3	50.73
24GHz	200	40	3	63.06
28GHz	200	40	3	64.4
38GHz	200	40	3	67.06
60GHz	200	40	3	71.02
70GHz	200	40	3	72.36

The reason I added the chart is to show the high loss you have at the higher bands. Now, while the number looks like it is double the loss, it is more than that. For every 3dB you lose 1.2 of your power, literally, lose half of your power.

Look at this chart showing power in Watts compared to power in dBm, pay attention! It helps put it into perspective. I like to look at watts because it is easy for the field engineer to see a loss in power.

Table 2

Power (dBm)	Power (W)
-30 dBm	0.0000010 W
-20 dBm	0.0000100 W
-10 dBm	0.0001000 W
0 dBm	0.0010000 W
1 dBm	0.0012589 W
2 dBm	0.0015849 W
3 dBm	0.0019953 W
6 dBm	0.0039811 W

9 dBm	0.0079433 W
10 dBm	0.0100000 W
20 dBm	0.1000000 W
30 dBm	1.0000000 W
40 dBm	10.0000000 W
50 dBm	100.0000000 W

We need to understand what the application will be for 5G to be successful. Let's face it; LTE would not have taken off without the demand for bandwidth. What caused demand for wireless bandwidth? You could say the laptop, but you know it's the smartphone, specifically the iPhone that changed the wireless world. Now we can't live without it, and even the president, (Obama) says it's a necessity. The link is at the bottom, so you know that I'm not making this up.

So how do we make money? The carriers will come up with something, but the mobility factors seem quite limited. The fixed wireless aspect makes a lot of sense to me, but if we can get 10Gbps to our mobile device, I am all for it, but who will pay for the backhaul? Latency is very important, so they must bring the computing to the edge, NFV and SDN will be a huge part of 5G. The latency should be low, and then the bandwidth will not be as critical, right?

What is the overall 5G plan?

Maybe this is where the devices will need to take a step forward again, like the iPhone. Then the networks pushed LTE out so that high-speed data could get the devices reliable. Then devices added more and more memory to improve it once again where the 5G miracle may happen. The devices will need to make another quantum leap. The way I see it is that the networks will dump data fast in large chunks for high bandwidth applications and the device will need to capture and process that data. The app will rest in the device, and the device must take the data and break it down. We are doing this now with applications in our smartphones and tablets, but they need to be the edge, they need to process it all quickly, and they need to be able to connect and accept data quickly and efficiently all while processing it for the application.

I really the think that the edge should push farther out so that the cloud will extend farther. Fog computing, where the edge is at the end, will push the network. Hop after hop, so that computing is close to the device or it may be on the device itself. Will there be awesome high-speed kick ass wireless devices that ten other devices can connect too? I hope so. Will fixed wireless take over the world? I hope so. What are you hoping for, other than a bunch more work in deployment?

One thing that could hold it back is that the carriers do NOT want to replace their networks. They have LTE in as the foundation. They do not want to do another forklift upgrade. They want to keep making minor updates until it completely maxed out. They are counting on LTE to push them well past 2020. When we all see 5G released in 2020, it will be on the back of LTE, which technically is still 4G but by then it should be on serious steroids.

One other thing, the carriers do not want to give up their dominance. They intend to rule the wireless world well into 2030. So, when the FCC has auctions, they will spend billions. Just look at how much spectrum AT&T is sitting on, you think they would roll it out soon.

Just a few ideas.

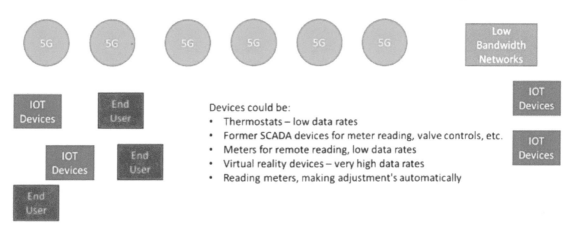

Figure 18

The 5G System

The system will be made up of most of the components that you see in 4G. I will go through it because some of the parts may be missing or change. Let's break it down.

Standard System

Just like 4G, it will be a mishmash of networks and formats. If you think 4G is LTE, then you need to go back and read over the beginning of the book again. When you think of 4G, you do think of LTE, but it is only the primary coverage option of this IP network. You have LTE and LTE-U and Wi-Fi all working together. What I'll do is cover the new format which is defined and how they intend to build the sites. It's being built and finalized. It will be part of the current LTE systems with upgrades.

Base Station

The BTS will be much like the existing system. They will have the RF transmit and receive just like current systems. There may be FDD and TDD systems, but I believe you will see more and more TDD systems so that operators can control the uplink and downlink bandwidths. There will be servers at the site that will help not only with traffic control but also with the cloud. The concept of 5G will be using the cloud more than ever. With that, they need to bring as much as they can to the BTS to improve latency for the end user. It is going to be a game changer that has little to do with the RF side of the house. There is more to this generation of communications than RF!

Antennas and Radio Heads

We have seen the marriage of the radio head and the antenna in the past from various vendors and OEMs. I know that T-Mobile worked with the all in one unit to try to save on tower rent. They deployed some of the equipment to reduce the number of jumpers and the real estate on the tower.

For 5G there are several advantages of combining the two into one. Antennas have come a long way, and now with MIMO happening, we will see even more connections between the radio head and the antenna. It may lead to more and more antennas that have the radio heads integrated. All in one unit would make sense, the connections between the two will increase. I think it makes sense to have the fiber and power all run to the antenna with no RF jumpers needed. The downside is the size and weight of the antenna. All that weight up the tower at the end of the mount. T-Mobile has been deploying this way at several sites, and they seem to be happy with the setup. I am not sure how tower owners will feel about it. Some carriers still need to ground mount because the tower can't handle the load or leasing restrictions.

How does MIMO work?

MIMO will be a key to getting higher throughput. Does 5G need MIMO, of course, it does, it is part of the equation? Look at MIMO at the way to get the signal to pass through multiple paths at the same time. When you see 4T4R, that means they have four transmit and four receive at one time. If it says 8T8R, then eight transmit and eight receive.

The antenna will need multiple connections to the radio head, one for each transmit or receive. If it's a TDD radio, then you just need the eight connections for the 8T8R.

The antenna itself will have many antennas inside of it. It is used for LTE and Wi-Fi all the time. It's where you increase throughput with the antenna by sending more than one signal to the antenna. Inside MIMO antenna you have several antennas all working simultaneously.

The only way it helps is when the UE has multiple antennas as well. While the 8T8R or 16T16R BTS can talk to a lesser UE, the bandwidth is limited. It's best to have the antenna setup match to both sides. MIMO helps with the more devices you talk to by using multipath signals to multiple antennas between devices. To get more technical detail click on the links below.

Research MIMO:

- http://tec.gov.in/pdf/Studypaper/Test%20Procedure%20EM%20Fields%20From%20BTS%20Antennae.pdf
- http://www.radio-electronics.com/info/antennas/mimo/formats-siso-simo-miso-mimo.php
- https://www.scribd.com/document/279219361/MIMO-Formats
- http://www.l-com.com/wireless-antenna-mimo-80211n-80211ac-antennas
- https://en.wikipedia.org/wiki/MIMO
- https://www.telcoantennas.com.au/site/how-does-mimo-work

Deploying 5G Small Cells

How will 5G small cells be deployed? How will 5G small cells be installed? How will the 5G small cell planning go? These are all good questions. Let's touch the surface to understand what the starting points are. Let's also cover how you can play a part in the 5G expansion.

I am not sure what the 5G format will look like in the mmwave spectrum, but we all know that LTE will be the foundation in existing spectrums. It could be LTE-Advanced or eLTE, evolved LTE. It looks like LTE will be here for a long time, in tech time anyway.

With the coming of 5G, we will see more and more HetNet. If you are in the industry, but the great thing is that the networks will be able to talk to each other. Much like the internet now, where you just plug in, the wireless networks will start to connect to other wireless networks. The carriers will be able to connect to an independent network and hand off data. They are doing it now with Wi-Fi. However, Wi-Fi is not as friendly with LTE as we would all like to think. That is why LTE-U will make things easier to interconnect wirelessly. How exciting is that? Make the carrier independent small cell a multi-carrier small cell. How cool is that? Looking at the unlicensed band as part of 5G is essential because many small cells in the unlicensed band will be used to offload the constrained licensed spectrum.

That is why you could build your network then reach out to the carriers to see if you could connect to them. The carriers made it clear that they don't want to pay for small cells or DAS unless they see a clear payback. However, I think they would entertain a partnership with a business or company that could help them serve their customers.

Just to note all the carrier say they are testing 5G, but it looks like Verizon and AT&T have made public progress here in the states. Verizon went so far as to set a standard, look for it here.

I don't' want to take anything away from T-Mobile, they tested 5G with Nokia just last month. You can read about it here.

The 5G Deployment Plan Handbook

First off, what is a 5G small cell? That is the question! If you follow me at all you know I say that 5G is the system, not necessarily the format Well, it is, but when you hear about the carriers talking about testing 5G, what does that mean? Let me explain.

First off, 5G will be a combination of LTE and new formats. When I talk about 5G small cells, I am talking about the 28GHz spectrum and up. Considering that licensed and unlicensed will be included. LTE-U and Wi-Fi will play a part. We need all the spectrum we can get. So, keep that in mind when focusing on the 5G small cell.

Now, the carriers are all testing 5G in small cells in the 28GHz spectrum and up. They are currently testing, in the USA, 28GHz because there is a lot of spectrum available there. The FCC approved it so let it be done! That is where most of the carriers have been testing equipment. Nokia and Ericsson are testing equipment in this range as well. So, that is where we will focus on today's discussion.

The small cell should have the basics, a type of BBU and a radio head. Only this time, at 28GHz the cable loss would be super-duper high! Sorry for the technical terms, you need a special cable to handle spectrum that high. They will have the radio head connected to the antenna. That can be done today, but most carriers, except T-Mobile, seem to like them separated so that can replace them if necessary. Usually, they wind up replacing both, but I am getting off topic.

The radio head and the antenna one unit, just like microwave where they attach the radio head directly to the antenna. Think about the extra weight of the antenna will increase. This could influence antenna placement and the way you mount the antenna. If it's outdoors, then you may need to worry about the weight or the size of the unit. Most times it may not matter. Indoors it may not make a difference, but you may not have the stealth unit you hoped for.

The BBU may all be in one unit for indoors or outdoors they could have it all inclusive or could be a separate unit with a fiber connection between it and the radio head. The expectation is that the radio head will have more intelligence in it so that the fiber runs can be longer for someone like Verizon who loves to deploy the CRAN system. It makes the core and BBU virtual. The radio head will be able to do more but won't have the full macro capability, but in theory, it should not need it. It should be able to see the BBU which may be nearby or in the cloud. How cool is that? Virtual radio heads are not far away, and the cloud will lower latency, but once again, I am getting off subject, let's focus on the small cell.

What has changed? It doesn't sound like much as far as the small cell is concerned. It sounds more like the same device, integrated into a different band that is smarter than the old device. So what?

Well, if you're in this business you should already know the answer! It's the antenna! That's right, the antenna and the connection between the radio head and the antenna! That is going to help us get maximum spectrum from the radio head to the antenna and from the antenna to the UE device! How? I am glad you asked!

MIMO, (Multiple In Multiple Out) which allows more spectrum to pass between devices. First off, the radio head will have 4 to 16, (maybe more) connections to the antenna allowing many bands to talk to the end device. Fun MIMO PDF to read here. It's getting highly technical, so bear with me. They cram a whole bunch of antenna panels into on antenna and then split signals to and from each little antenna. But wait, there is more! Now the little antennas can focus on a specific user giving them a priority and allowing it to pass more and more traffic. OK, maybe my explanation wasn't so technical, but NI.com does a good job explaining massive MIMO here. My point is that the antennas are way smarter and they will be the game changer here.

MIMO can be used on a macro or small cell, but for small cell, the additional spectrum will help because we have huge bands at 28GHz and up. In fact, Sprint has a ton of spectrum at 2.5GHz and up, but they seem to be very slow to deploy. Maybe they would sell it; I have to think that they would because the CFO recently said how the spectrum would be perfect for 5G. Antennas got smart so that it will help you will help you with throughput and a focused signal to the UE.

Massive MIMO is going to make wireless connectivity better and better. Who knows what we will see next. I want to see this technology rolled out. It can be used in most every band to improve throughput and coverage to the device. Think about what the cell will do and how the devices will need to add more antennas to utilize it. If the antenna can have one smaller antenna talk to a specific device while talking to other devices simultaneously using common spectrum, it will be a game changer. I know there is more to it than that, but it sounds cool.

The antenna technology will take more RF engineering; I believe because planning will take more time on smaller networks. On the other hand, they will be in a higher band, 28GHz and up with more spectrum allowing the installer to install based on LOS and then they can use the spectrum planning to keep the channels from stepping on each other, just a thought. It would make a great neutral host. We thought Wi-Fi could do that, but it never worked out. These are all thoughts that I am sure someone is smarter than I will finish.

There is a nice PDF found here, http://telecoms.com/wp-content/blogs.dir/1/files/2016/06/5G-architecture-options.pdf that covers the architecture of the 5G radio. Please note that they clearly talk about technologies, but the reality is that it will include both LTE (which they will call ELTE) and the NR, (New Radio developed by Qualcomm). We can talk about that in a future blog if you're interested.

FYI – 28GHz will be licensed, but I am not sure how. I say it like that because the coverage area must be very small. I would like to see it as lightly licensed. I see it used more like the 3.65GHz band, only smaller coverage area or as a backhaul. With all that spectrum, you could have backhaul depending on the usage. More info here.

Resources:

- http://telecoms.com/wp-content/blogs.dir/1/files/2016/06/5G-architecture-options.pdf
- http://www.slideshare.net/Netmanias/3gppr3-161772att5gmigration160827141718
- http://4g-portal.com/tag/elte
- https://www.wirelessweek.com/news/2016/07/fcc-unanimously-opens-nearly-11-ghz-spectrum-5g
- https://www.ptc.org/assets/uploads/papers/ptc16/MON%20KN_Onoe%20Seizo.pdf
- http://www.rcrwireless.com/20160714/carriers/verizon-att-t-mobile-sprint-deep-5g-testing-across-high-band-spectrum-tag2
- http://www.rcrwireless.com/20160711/carriers/verizon-completes-5g-radio-specification-tag17
- http://www.5gtf.org/
- http://www.ni.com/white-paper/52382/en/
- http://www.ijergs.org/files/documents/A-NETWORK-42.pdf

Related Blogs:

- https://wade4wireless.com/2016/10/11/deploying-5g-small-cells/

Will 5G be a Success?

Will it succeed? Of course, the issue that I see right now is not the technology, the loading, the multiple access, but the spectrum. How far can millimeter waves go? How many connections can 28GHz or whatever band they're in, can it serve? How large of an area can it serve? We all look too small cells or CRAN to perform here, but will there be a payback. Can the carriers make a play for fixed wireless or will they try to capitalize on dense networks? Can they make the dense network business model work? Does it have to be a carrier to build this or can it be a smaller business who could tie back into another company's core? Will they need a mobile core or can it start to replace fiber? Remember, there were so many microwave hops connecting broadband when fiber was not so readily available.

I don't think that the question of 5G being a success is fair. It's more like, "Will Millimeter Wave be a failure?" While it's only a part of the overall picture, we all know the rest are successful. Isn't that the big question here? The elephant in the room, so to speak. Can mmwave perform the way that today's spectrum does? Can it go beyond PTP?

The real winner will be the fiber backhaul companies; they will offer maximum bandwidth, chances are it will be dark fiber dedicated to the carriers, one more expense the carriers to look at before deployment. It's the big cost. Sorry cable companies, but the cable modem may not be able to service such a huge capacity. If they do virtual reality, they will need to get to the edge for low latency and have a lot of bandwidth and the devices will need to maintain connectivity. Or will they?

So, to sum this up, I hope it is a success, for the entire wireless industry. Whether it uses licensed or unlicensed spectrum, I see an opportunity for all.

The 5G HetNet

One thing that we will see is that the network will encompass more than a macro or small cell network. It will be everything we need wirelessly. Most of the 5G will be the smaller networks since they intend use millimeter wave in the Extremely high-frequency band. It's going to have many limitations as I listed above. The mmwave network will rely more on the smaller cells and possibly digital DAS. It is going to be a combination of BTS units, probably mostly small cells, and it will target specific customers and applications. Macro sites may do more than just provide mobility; they will feed small cells and CRAN systems.

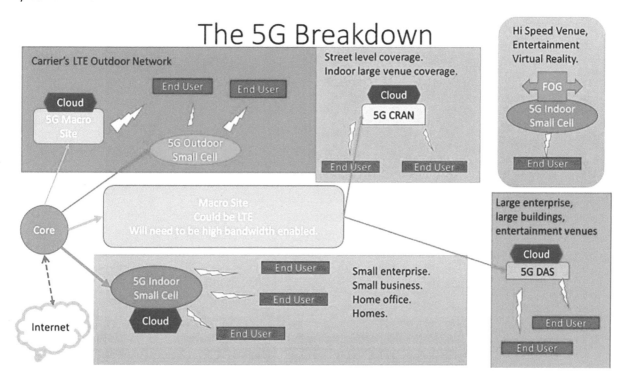

Figure 19

The new network will resemble the old network when it comes to deployment. I think it will be good to know that it's a HetNet from the beginning and it will take several different types of BTS as well as technologies to get us there. Let's not forget the mix of the spectrum, licensed and unlicensed that will be required to get the signal to the end user.

The Cloud RAN

I mentioned CRAN earlier. It will play a big part in the future of networks because it will take out the BBU. The goal is to use as little equipment to use as little equipment at the site as possible. If you can get away with a router and a radio head at the site, whether it's a small cell or a macro site, then you have something that could be a viable solution with less rent. I believe that small cells are already there.

What you need it to have a great backhaul with low latency. There it is again, low latency. It appears that to have this work properly we need bandwidth *and* low latency, it can't be one or the other. We learned this in the fronthaul applications. The radio head has timing issues that are critical to the use.

The OWM needs to eliminate the timing issue, or the carrier should get the BBU functions close to the radio hear.

CRAN is being worked on and tested in China. They seem to have some working models, but it hasn't caught on in the states just yet.

What is Edge and FOG Computing?

Edge computing is pushing the applications away from the centralized server to be processed beyond the cloud as close to the end user as possible. It's going to help keep latency down in the future. It will not rely just on the cloud but on a server, in this case in the cell site, to serve as an edge computer. I believe this will be a key factor for IOT, controlling and monitoring devices with very low latency. The apps to run them could be very close to them as the data is sent to a central server to collect data about the thing.

Fog computing was a term created by Cisco, that seems to do what edge computing does. A cool term that takes the apps beyond the cloud and as close to the end user as possible. From what I can tell, Cisco came up with this especially for IOT to bring the application as close to the devices as possible. They looked at it as pulling the cloud as far out to the end user as possible. Great concept and one that can be used today. Thank you, Cisco.

What is different between the 2? Apparently, fog computing is more scalable per the article from readwrite.com which I have a link below. From what the article in Automation World, link below, Fog computing pushed the intelligence down to the local network whereas Edge computing uses the edge appliance to push the data to the device that can hold it. They use the example of a PAC, Programmable Automation Controller.

- http://robtiffany.com/the-cloud-is-dead-long-live-the-edge/
- https://en.wikipedia.org/wiki/Edge_computing
- http://www.webopedia.com/TERM/F/fog-computing.html
- https://www.cisco.com/c/dam/en_us/solutions/trends/iot/docs/computing-overview.pdf
- http://www.howtogeek.com/185876/what-is-fog-computing/
- http://readwrite.com/2016/08/05/fog-computing-different-edge-computing-pl1/
- http://www.automationworld.com/fog-computing-vs-edge-computing-whats-difference

What is SDN and NFV?

To cover SDN, I have a few links in here that may help you. On the wireless side, you won't worry about SDN or NFV. They are a router and network function.

SDN = Software defined networking.

NFV – Network Function Virtualization.

SDN is software defined networking, and routers will handle it. It is evolving every day. The thing to remember is that it will make networking very flexible and automated. It will be the tool in routers that allow the routing to be more efficient. Through SDN the cloud will become easier to use because the complex routing is made easier and automatic. It will add consistent policies to the network. It will scale quickly and efficiently. Network policies become simplified.

As for NFV, it takes the work from the IT guru and puts it in the routers and firewalls. Now they can work on network policies with minimal guidance for the IT groups. The functions are done virtually inside the equipment. A game changer for the network making functions happens virtually without constant human intervention.

The reason SDN and NFV are used together is because they happen to merge the network, making it smarter, so work's done virtually.

If you would like to learn more, the links below are a good start. The rest is up to you.

- https://www.cisco.com/c/dam/en/us/solutions/collateral/data-center-virtualization/application-centric-infrastructure/sdnfordummies.pdf
- http://packetlife.net/blog/2013/may/2/what-hell-sdn/
- http://www.infoworld.com/article/2606200/networking/111753-Software-defined-networking-explained.html
- https://www.sdxcentral.com/nfv/definitions/whats-network-functions-virtualization-nfv/
- https://www.mushroomnetworks.com/blog/2016/08/02/what-is-network-function-virtualization-nfv-and-how-will-it-impact-your-wan/

What about Wi-Fi?

First off, Wi-Fi will be part of the ecosystem. If LTE-U can take off, then I believe that carriers will make a run for LTE-U because of the operability compatibility with licensed LTE. It hands off so much better than Wi-Fi.

I never rule out Wi-Fi as a critical part of the ecosystem, but I also know it has many limitations and I know that the general attitude of people in the USA is that Wi-Fi should be free. Like it or not, that is what most people think of when they think of Wi-Fi. When you use it in a public place, you general have problems if there are multiple hotspots. That's been my experience.

I have done many Wi-Fi deployments ion the past. They are a common deployment where it can be rolled out very cheap because its license free and the hotspots are very low cost. A company called Ubiquity has flooded the market with very inexpensive equipment that many people have deployed and carrier grade Wi-Fi vendors, like Ruckus, have a hard time pushing their carrier grade equipment even though it is very high quality.

The US carriers, specifically AT&T and T-Mobile, are big fans of Wi-Fi but even they have had problems using it nationwide. They still prefer the licensed spectrum for much of their traffic. They do see value in Wi-Fi, but it is merely an extension of the licensed spectrum that they paid billions for.

However, carrier-grade Wi-Fi needs the following:

The WBA Guidelines state that the following standardized carrier-grade Wi-Fi capabilities are needed

to ensure networks can scale to meet the requirements of the industry:

- Consistent User Experience
- Network Discovery and Access
- Authentication and Security
- Service Experience
- Fully Integrated End-to-End Network
- Network Architecture
- End-to-End Service Provisioning
- Network Management
- Network Quality
- Network Security
- Network Manageability

I didn't make that up; I got it from a report called "From 2016 to 5G, Wireless Broadband Alliance Industry Report" which is put out by Maravedis-Rethink, (http://shop.maravedis-bwa.com/products/from-2016-to-5g-wireless-broadband-alliance-industry-report) a wireless infrastructure analyst firm. The WBA, Wireless Broadband Associates, is a group of Wi-Fi professionals, learn more about them at www.linkedin.com/company/wireless-broadband-alliance. They do all that they can to promote Wi-Fi, which is a great format and has been around for years. Their dream is to see more and more Wi-Fi go carrier grade. They see LTE-U as a threat, and in my opinion, they have done all that they could to delay the release of it. You see, they were hoping to get the carriers to pick up more and more Wi-Fi carrier grade systems. While the carriers do like Wi-Fi because it is a great way to offload the data, they never had any interest to adopt it as the license-free format of choice.

To deploy Wi-Fi, you need to have several things in place. You still need backhaul. You still need the mounting asset. It is comparable to a small cell installation. So, let's go through what you need for Wi-Fi.

Cheap and Dirty

Many companies still do RF design, but the reality is there is Wi-Fi everywhere. In many cases, an interference study is a waste of time. It's probably already there. Wi-Fi is a "best effort." The best you can hope for is to align your channels the best you can without getting interference from the next guy. Most devices have built-in analyzers, so many deployment crews just check it out after they install the equipment. Is this the best practice? No, but it's cheap and effective.

Mounting it is straightforward, and it may depend on what area you want to cover. If you're indoors, you need to know what the loading will be for good coverage.

Remember that you still need backhaul and chances are good if you are doing a large deployment you need a management system, Some Wi-Fi carriers just put out the hotspots and have a server to let their customers log on. Many just have something to allow free access for anyone who wants to go on. They still have some access for people to log on and accept the usage terms.

A straightforward and easy way to deploy Wi-Fi. Many of these operators make deals with local building owners, businesses, or even local governments to use a shared backhaul to serve their customer base. Once again, the backhaul is where most of the money is going. Whether it's fiber or copper or a cable modem or even a T1, it is a monthly reoccurring cost that you must pay.

Carrier Grade

Now there are exceptions to the cheap and dirty way I talk about up top. For instance, in sports stadiums, and in convention centers you need to do a lot more planning. Many of those venues need to know the traffic flow and usage at the peak times so they can plan to have a limited number of users on a hotspot at one time. Here is where planning and design play a huge part in the deployment. If you look at these venues, they have carefully planned the location of each hotspot and antenna to serve the customers with the best efficiency. Since Wi-Fi hotspots and antennas are relatively inexpensive, they know that in these venues they can apply them sparingly. Here is where the backhaul will cost the most money.

When dealing with this type of deployment, you need to have a great management system. The Wi-Fi hotspots need to be able to handoff properly. It helps you decide which vendor is the best fit. It doesn't pay to grab the cheapest equipment unit and hope for the best unless you have a small lightly loaded system. You want a quality gear that is carrier grade made by a reputable OEM for the Wi-Fi. You will have more options and an ecosystem for control. Remember that you want to have a management system and control that wouldn't normally be available with cheaper vendors. Make sure you plan this out and think it through. Also, go with someone who has been around for a while. It pays to do your research.

The other thing for these venues is that the RF design is crucial. You want to know what the RF coverage. I wouldn't worry as much about interference, but I would worry about what area the antennas will cover. I would also want a channel plan laid out so that we don't have self-interference. Whether it's 2.4GHz or 5.8GHz, you want to be sure you have it planned. Even with the plan, you may have issues with self-interference. RF design for this type of venue is critical.

Another thing to think about is what apps you want to use. I know stadiums like to enhance the user experience, so if they were to run video apps that promote the game or event in the stadium, something that someone can't experience at home. Maybe at a convention center, they want to have a map or an internal tracker of where and how many people are passing by each booth at a show. It needs to be considered up front.

Do the proper planning up front; then you will be better off in the end. Chances are good you could miss something. Each thing that you overlook will have ten things that you planned. The odds are in your favor.

Who will win in 5G?

We have all heard about 5G, the next big thing, for the carriers anyway. Will it be all that we think it can be? Super high bandwidth with awesome applications? Who will win the 5G race?

The way I see it Verizon Wireless, and AT&T will win, or at least be there first. They invested heavy, they are working to put out real standards, and they want to win. The other carriers seem to be playing with but not taking as seriously. Sure, they all say they are working on 5G, and I am sure they have something in a lab with high bandwidth in the 28GHz range, or somewhere up there, to test these applications. That way the investors are happy along and the public is impressed. I mean we all want to see virtual reality happen very soon from a wireless device, right?

Verizon seems to be working hard on the technical standards and the testing of the network. They are breaking ground to make sure there are standards in place. They see the need for speed and new applications. They are known not only for being on the cutting edge and the better network.

AT&T seems to be finding ways to use it and testing it as well. They know that they can build better technology for automobiles and drones. Something that AT&T has been very active in, and they intend to build on what they must improve the quality of experience. They know that they can improve the network and speeds to the end user.

While I see that Sprint and T-Mobile are looking at it and testing in labs, they probably are in no hurry to lay out a lot of money just yet. They may wait to see what standard will be adopted then take the money they saved and use it to build a network or expand what they have. I hope they are planning for this future. In all honesty, T-Mobile is doing a great job of laying the groundwork for high speeds while Sprint, well, Sprint doesn't seem to be doing much of anything for the future.

The real winners should be the backhaul suppliers and the carriers. They find a way to build networks for support that the customers want. They will get reoccurring revenue for the network and equipment, along with the landlords that rent the space to the carriers. They will get income for this that will be ongoing. It is the one constant that they need to provide something that is needed and get revenue for it until the system changes or moves.

The winners in 5G will be the end user, but let's all try to build out the best network available. They will expect to have a great quality of experience. They expect to have a new application that can run on 5G. They will want to have the bragging rights. However, it should work, or the reputation of 5G will suffer immediately. People expect reliability as well as the coolness. They want to have something that will work every time. I know, if you have a smartphone that not every application works. Something that alleviates the carrier's responsibility when they provide the connectivity that's expected. Not only that but the backhaul for 5G will need to be better than ever. So, this will matter quite a bit. If we have 5Gbps over the air but only have a backhaul of 100Mbps, then it will suck and we will complain, but will the end user know who to blame? I believe they will figure it out. Just like the location of the cell, it should be at a good location that no one can see.

All of this will play into the winners and loser. I believe that T-Mobile will be smart enough to let the big boys deploy and learn from their mistakes, they seem to pay attention to what they do so they can see what works and what doesn't work.

The Real 5G Winners Will have VISION!

If the FCC doesn't box out the little guy, then you, the small business owner will be the real winner. Seriously! The end user will also be the real winner. You may have already figured this out, but let's talk about it anyway.

The end user will have many options for wireless. They will have the opportunity to use so many functions that were not available before. Artificial intelligence will go to new levels no matter where you are. Virtual reality and augmented reality will become an awesome reality. They are going to change the way we live life. We will have new opportunities to do new services that we've never seen it before.

With the new services, many businesses can get their 5G spectrum. I think that mmwave will be something that most companies or people can lightly license. Then they can have a private virtual reality system. Amazing for entertainment. It will be a game changer.

What about having your IOT system? You could control or monitor your devices in the building or on a campus with your spectrum. Monitor what you want with your spectrum, don't pay a carrier! Pay the FCC for a lightly licensed chunk of spectrum.

Take advantage of the new spectrum that will be available. New devices, new technologies, and new ways to take advantage of new opportunities. Come up with new business cases and plan to do something amazing!

The 5G Deployment Plan Handbook

Resources:

- http://www.phonescoop.com/articles/article.php?a=187&p=232
- http://www.rcrwireless.com/20160622/carriers/researcher-internet-things-wont-make-operators-money-tag17
- http://www.npr.org/sections/thetwo-way/2015/01/14/377230778/obama-pushes-fcc-to-expand-broadband-access
- https://www.youtube.com/watch?v=ikR0_ptc4P4
- http://www.rapidtables.com/convert/power/dBm_to_Watt.htm
- https://www.pasternack.com/t-calculator-fspl.aspx
- http://www.qsl.net/pa2ohh/jsffield.htm
- http://faculty.poly.edu/~tsr/mmwave.php
- https://www.youtube.com/watch?v=pN_3Iek2jNw
- http://ieeexplore.ieee.org/stamp/stamp.jsp?tp=&arnumber=5590362&tp=?ALU=LU1029396
- http://faculty.poly.edu/~tsr/Publications/icc2013.pdf
- https://www.microsoft.com/en-us/research/wp-content/uploads/2015/03/Xinyu-Zhang_5GmmWave.pdf
- http://users.ece.utexas.edu/~rheath/presentations/2015/mmWaveFor5GTWS2015Heath.pdf
- http://users.ece.utexas.edu/~rheath/presentations/2015/MillimeterWaveAsTheFutureOf5G_5GForum2015Heath
- http://www.fiercewireless.com/tech/at-t-seeing-14-gbps-to-one-user-5-gbps-to-two-users-5g-tests?utm_medium=nl&utm_source=internal&mkt_tok=eyJpIjoiTm1GbFpqQmlORE0zWWpBeClsInQiOiJmNjRSNnpTUzhhNWlvSmlxVzhWMkF6YWYxMlo5emw4alFRTVJXbG9rdnAyMDZ1WG01UVhRdzd4N2NhZXJpRkxrV2VFcm5xTm9Lb0NmRWVsXC9za0tNVDMwejRiamUwU0JwR3dnNnNFRFlmcHc9In0%3D
- http://www.rcrwireless.com/20160812/carriers/small-cell-mmwave-lte-unlicensed-tag17
- http://www.lprs.co.uk/assets/media/Rethink%20IoT%20Wireless%20Market%20overview.pdf
- http://www.pcworld.com/article/2985233/internet-of-things/lte-standard-for-machines-gets-the-green-light.html
- http://www.3gpp.org/news-events/3gpp-news/1607-iot
- http://www.wired.com/2016/01/wifi-halow-internet-of-things/
- http://www.techrepublic.com/article/802-11ah-wi-fi-protocol-for-iot-solves-two-m2m-problems/
- http://electronicdesign.com/iot/understanding-protocols-behind-internet-things
- http://iotpodcast.com/
- http://pages.silabs.com/rs/silabs/images/Wireless-Connectivity-for-IoT.pdf?mkt_tok=3RkMMJWWfF9wsRoguKjNZKXonjHpfsX86%2B4rWKK3lMI%2F0ER3fOvrPUfGjl4DSsJkI%2BSLDwEYGJlv6SgFTLPBMbNsz7gOXBg%3D
- http://postscapes.com/internet-of-things-protocols/
- https://en.wikipedia.org/wiki/LPWAN
- http://www.semtech.com/wireless-rf/internet-of-things/
- https://www.micrium.com/iot/devices/
- http://www.networkcomputing.com/internet-things/10-leaders-internet-things-infrastructure/1612927605

- https://www.thethingsnetwork.org/
- https://www.youtube.com/watch?v=2nsEAw_SirQ
- http://cdn2.hubspot.net/hub/213677/file-2170475006-pdf/Five_steps_to_get_a_Public_Safety_4_9Ghz_FCC_license.pdf
- https://www.fcc.gov/49-ghz-public-safety-spectrum

The 5G Business Case Foundation

What is your Business Case for Wireless Coverage?

What is your business case? What I mean by this is what your business is? I am going to provide some examples of business cases below based on your needs and business. Remember that I said to start with the end in mind, here is the foundation.

Could be for 5G or Wi-Fi or LTE-U or anything that you may have a need for in your business. The reason this will be so relevant with 5G is that the networks can be connected externally to the Internet or another wireless network.

When you go through the list, you may think you need "licensed" spectrum for all the business cases. You don't. You don't necessarily have to spend billions like the large carriers did you get the coverage you need. You could go unlicensed, or you could go lightly licensed, you could lease spectrum for the carrier or another company that has some. I want you to eliminate the excuses. Stop thinking that only the carriers have 5G spectrum and that only the carriers rule the airwaves. You have the power to build your system for your need.

When you look at what you need to build the system, this is a high level only. To drill down into the needs of your system, you need to look at quite a bit of detail, like backhaul and fronthaul. Use fiber or wireless or copper or a combination. You will need to pay to build it and maintain it. You may need to worry about repair and support. What about the upgrades? It takes money and support to keep everything up to date. Think how many times your smartphone updates, your system will need to be maintained and kept up to date. It doesn't need to be too complicated, but it can be.

It may change from business to business, but generally, I go in this order:

1) Building the business case notes.

2) Coverage – you need to know what coverage you need.

3) Goal and growth – what is the goal of the system

4) Budget – you now have the information to build a budget.

5) Spectrum – what would be the recommended spectrum?

Medical and Health Care

If you are in health care, then you probably want to cover a hospital or a medical center for a specific reason. Your customer may be the patients or doctors or specialists or technicians. For patients, it may be a public network, but are you going to build a private network.

1) Build your business case so that you can build your budget. You can start construction of the system with a solid plan in mind. It will be more than a wireless network; it will be an essential lifeline for many. Remember that privacy matters, read more on privacy in Wi-Fi at http://corporate.findlaw.com/litigation-disputes/hipaa-and-wifi-regulatory-tangles-for-wireless-health-care.html to see how it might affect any system you install. Also, go to https://meraki.cisco.com/lib/pdf/meraki_whitepaper_HIPAA.pdf and

The 5G Deployment Plan Handbook

https://www.allpointcompliance.com/Blog/19159/HIPAA-and-Wireless-Security?_escaped_fragment_=#! To make sure you are compliant.

  a. Hospital coverage for patients and visitors. Wi-Fi coverage and cell coverage would be just what they need to help the time pass by and to alert the family of the ongoing conditions of the patients.

  b. Paramedics coverage so they can collect data before they arrive at the hospital to treat patients properly. They would know the allergies before reaching any hospital and looking up medical records. They could see it all on the scene within minutes. Just imagine if they could scan someone's fingerprints to get their medical information immediately. All the health records available at their fingertips, pardon the pun.

  c. Emergency care centers and hospitals could give their doctors instant access to data on a tablet anywhere in the hospital. I believe they have this capability now, but if they could get immediate updates or even look at an updated diagnosis immediately after it happened, there would be no delay in changing treatment if needed.

  d. The remote diagnosis could happen with accuracy. How, think of virtual reality and how you could do a brain scan, MRI, or a live scan on a patient and someone, a specialist somewhere else in the world could look at it real time using virtual reality seeing what the machine sees in real time. They could not only see it but also look at it in all angles, maybe even control the machine doing the scan and talk to the patient at the same time. Saving someone's life by making specialists available anywhere from one location!

  e. Remote surgeries could happen with robotics and massive bandwidth. How? By having a remote doctor look and control the robot along with the local physician to help the patient along. They could upload surgeries for someone halfway around the world, looking at the live video while the surgery is happening. Taking the time to make evaluations and talking to the local surgeon while the specialist is operating the robotic surgical apparatus all while the patient is receiving the best care possible. A new perspective to doctors without borders! WOW! 5G is a game changer for medicine worldwide!

2) Coverage - Cover your building (or campus), which could be:

  a. Hospital – I have covered hospitals, and the things to look for are crazy. They have lead lined rooms for X-Ray which means no RF will penetrate. Do you declare it a dead zone? Probably, but the real issue is that you need to make sure you have antennas on each side of it, around it. The room causes problems all around it. Make sure to do your due diligence when covering such a complex building. Also, remember that there are lines run through it, like oxygen. It's more than plumbing and electrical in a hospital. By the way, oxygen can cause explosions.

  b. Medical Centers can be just as complex as hospitals but on a smaller scale. It is a good idea go not only have the plans but to talk to the doctors and maintenance group to see what is really in the building. They don't always document what is there, even though they are supposed to. What I am saying is, don't just trust the drawings, ask around.

c. Emergency Care Center is a place that you may think is like a hospital. They are an extension of the hospital with all the things a hospital may have but on a smaller scale. Plan carefully.

d. Surgery Centers prep for planned surgeries only. It has been my experience that these buildings plan very carefully. I have had a lot of luck with drawings for these places, but again, ask around.

e. Patient Care centers, generally doctor offices. They should be straight forward to cover.

f. Administrative Offices that are for the paperwork only. Another straightforward deployment.

g. Medical campus or hospital grounds could be an issue. Here is an area where you may have to cover outside of the building. Now you could have issues with privacy. When I was doing wireless backbones for hospitals and medical, they were very concerned about privacy. There are rules under HIPAA for the privacy of the patients to be protected at all costs. Think about the coverage. If it is simple to access Wi-Fi for the patients, then it will be handled differently than if it's the coverage for hospital staff. Medical records should are treated with the utmost confidentiality.

h. Mobile coverage, this is handled by a carrier, most medical companies don't spring for too much outside the building or off the campus.

3) Goal - What is your primary goal?

a. Track equipment, equipment the hospital gets lost or misplaced all the time, so why not track it. Make your hospital a smart building to locate equipment quickly on a computer or tablet with an app. RFID tags on the equipment.

b. Page doctors for emergencies, (obviously).

c. Update patient records. Use tablets to keep records up to date then if someone sets a tablet down, use a location app to find it quickly. Patient records will be available to many nurses with proper privileges. Quickly, easily access, and anywhere in the hospital or just in a zone that is approved.

d. Notify an alert when equipment leaves a section of the hospital. Prevents theft of not only equipment but data if a tablet is stolen from a specific area like intensive care. HIPAA is a serious thing! (HIPAA is the Health Insurance Portability and Accountability Act that protect patient's privacy. If you have ever done wireless work for hospitals, then you are aware of the HIPAA.)

e. Connect all the admin computers with a secure and private connection to avoid running wires everywhere.

f. Internal wireless phone system so that all staff members can communicate anywhere on a floor or in the hospital. They should have devices that would have all the necessary features a Dr would want.

g. Patient monitoring and tracking so that patients can be moved and tracked in the hospital or medical center while their vitals are monitored. If someone is going into cardiac arrest in a hallway, then the alert will come through immediately identifying the problem and the location.

h. Connect paramedics to the hospital when they are on their way there with a patient. Again, mobile coverage but this is where they could start treating the patient in the vehicle. If they could have all the details before getting to the hospital, then they would be aware of any allergies that could kill the patient. They would be so much more effective if they have the information ahead of time.

i. Remote countries could have robotic surgeries or training using high bandwidth. How? It would be in a hospital, and the data would be sent in from someplace on the other side of the world. A robot could perform surgery. Then, if done right, MRIs or brain scans or any scan could be seen by someone in another country where they have the expertise for that symptom, and they could make a diagnosis using 3D coverage. If the bandwidth is there, in that country and the specialist is in another country, he can help remotely with all the data. He could even remotely perform the surgery using robotics in another county by showing it what to do. I know this involves virtual reality and artificial intelligence to a certain point, but we are almost there! Isn't it exciting that we can help more people around the world?

4) What is your budget? You will need to decide how much you want to spend before you determine what equipment you will purchase. Limiting factor if you are starting out with a preset amount. What most companies do is determine the need then ask an engineer or solution architect to create a design, which will cost money, then they know what the budgetary numbers are for the equipment and installation and testing. Don't forget that you need to purchase the devices. Many people design the system but don't add the cost of the end device, and they cost money too!

    a. Once the use is settled on, ask for a system design.

    b. Determine how many user devices you will need.

    c. Determine your backhaul needs and the cost to support it.

    d. Determine your estimated usage.

    e. Determine what support personnel you will need to support and maintain the system.

    f. Determine what the warranty is what the software updates will cost.

    g. Determine the life cycle of the system, technology moves quickly, will you grow or replace the system in 5 years?

5) The spectrum needs to be thought out by where and how it will be used. Bandwidth could be critical, or low latency could be critical.

    a. "In building coverage" is obvious to provide internet access to patients. Wi-Fi or LTE-U would be the ideal spectrum for much of this use because it is license free, which is free and available on smartphones! They may want to use a licensed band, but most medical centers leave that up to a carrier to cover the med center. If they decide to partner with a carrier, then it is up to the carrier.

b. In building for special services, like robotic operations by remote doctors or remote doctors giving their opinion for treatment while looking at video or MRIs of a patient while it is happening! These applications have very high bandwidth needs and would need to have mmwave to be effective. That way the doctors could see what is happening real time, full video. Latency would depend on where in the world they were, but with the advances in robotics, remote doctors could guide the robot while seeing the patient real time during the operation.

c. Campus outdoor coverage would also be something like Wi-Fi or LTE-U. It would be something very cost effective with off the shelf hardware to deploy. All the money will be going into the security of the information. Network security will matter more than spectrum and will get a sizable budget towards it. HIPAA regulations will require heavy network security before a patient's records travel over it. I dealt with that when working with a point to point microwave to connect hospitals. We set up a remote MRI monitoring service and security on the network and over wireless was taken very seriously. HIPAA compliance was critical.

d. For mobile coverage, probably a carrier. I don't see any medical budget allowing for anything outside of the building or campus. I am realistic here; they would use an existing service whether it is a carrier or a public safety partner. I would think public safety, like FirstNet, would allow them to use their network for a fee.

Utilities

The Utility may want to have their network for several reasons. They may want to read all their meters on a private network. They may want to have employee connectivity anywhere they travel. They may want to upload real-time data of an outage or emergency in their area. They may remotely monitor equipment, could be power lines, gas or water gauges, or security in a remote location. Traditionally done with SCADA systems, but in today's world of IP, it could be better and done more efficiently. However, SCADA systems were very secure, and we all know that IP systems could be hacked. That is one reason that utilities may want a private system, for better security.

The downside, of course, is that it is very expensive to build out a network. The equipment and installation will be quite a bit of money, but the site rental and maintenance and the backhaul may be too cost inhibitive to build. In the past utilities started to build out WiMAX systems only to have problems with OEMs, connectivity, and reliability. They got a taste of the costs involved in building a wireless network. However, many switched to a carrier only to realize that the monthly costs add up and they have issues with keeping the costs in line with their budgets. It will be a never-ending monthly bill that only goes up, never down.

Let's look at what we need to build a business case for a utility.

1) Build that business case for your needs. Utility solutions have more to do with what the job is. Meter reading will be an IOT system with low bandwidth, battery saving, wide area coverage.

Substation and building connections are another matter because they may just collect data or give the people inside of those buildings access to the internet. Most utilities will need several systems for several purposes. What is the purpose of that specific system? Don't try to multi-purpose the system! Meter reading should be dedicated!

2) Coverage – what are you covering?

   a. Meters - Do you have a geographical area to cover meters? If so, then you need to think about how to cover them because many meters are in residential neighborhoods. Most neighborhoods hate towers and may not have poles. It is a serious consideration today because the higher bands do not cover as well as the old bands.

   b. Campus or large building – many utilities have a very large building where they monitor the status of meters or pressure or anything that if there is an issue an alarm needs to sound and the equipment needs to be monitored. An ideal case for a private network where you may have a firewall to the outside world but inside the equipment is on a private network without coverage issues. I have seen these networks built on Wi-Fi and they work. They do not need a lot of bandwidth, but they do need to be connected reliable and 24/7.

   c. Building – many buildings may need admin coverage for these networks to work reliably. Here's a case where you may want to move admin people around, or you have technicians you want them to connect freely without running up data costs on a carrier's network.

   d. Lines – by this, I mean gas lines or wire lines or water lines. A wide area of coverage.

3) Goal – what is the goal?

   a. Meters – this is a low bandwidth application, but it's one that needs a wide area coverage and could cost a lot to deploy. You may need to pay tower or pole rental charges.

   b. Tech laptops – if you want to connect the tech laptops for them to have remote coverage, then they will need to have enough data to perform their tasks, like remotely read meters, download email and manuals, IM back to the remote technical support center. Coverage is an issue as well.

   c. Monitor gauges – this could be set in place to make sure that the gauges or high voltage lines or high-pressure gas/water lines are all operating properly. Your first alert system. I would imagine that you would place the meters strategically around your region. Outside, in a building, on a campus, or even underground. How will you connect this meter back to your NOC?

   d. Admin usage – this is generally in a building as a convenience to your people working in the offices. It makes it easy for them to move around the office to work.

e. Monitor test equipment – this is an application where you could tag your test equipment or tools or spares so that you can track them wirelessly in a building, campus, or region.

f. Usage – this is where we monitor usage with sensors or meters for a specific pipe or cable. Monitoring mains, electric or water or water or gas. Making sure that the heavily used pipes are monitored 100% of the time is just smart. Here's a way to set up remote monitoring that can be accessed anywhere.

4) Budget – now that you have the details worked out, build the budget. Think about the end device, like the wireless device in the meter, the battery life of the device in a meter if you don't have a source. How often you plan to read the device like once a day or once a week or once a month. It will change the equipment you need and possibly the installation of the equipment. Think about how you will deploy the remote devices, will you need to pay rent, will you need to mount on towers or poles? It all adds up. What about maintenance, support, updating security, spares, etc. These are all issues that will change your budget numbers. Will you support the system internally or externally? What about backhaul at each site? What about the servers at the central location?

5) Spectrum is going to vary.

a. Indoor could be something simple like Wi-Fi or LTE-U, simple and cheap and cost effective. The exception would be if they were running high bandwidth applications, which I think is unlikely.

b. Meter coverage may be through a carrier or an IOT system, like NB-IOT which may be in the ISM band, like 915MHz or it might be LTE if they can improve the battery life.

c. Fixed wireless to buildings could be 3.5GHz or something like that. They would want some decent bandwidth to connect, but it would also be something lightly licensed. I think that most utilities would use a carrier for something like that unless they wanted their network for privacy and security. Also, they would have lower latency. Another alternative is the mmwave. Offering connectivity from nearby fiber to a substation. What I have seen is that most substations don't need lots of bandwidth, but it would work, and it could be a cost-effective way to connect the buildings given the proper circumstances. By circumstances, I mean line of site and weather conditions and surrounding vegetation. All the variables in any wireless system.

d. For mobile coverage, they may use their bands, but it is expensive to build and maintain. Today's carriers still use LMR, but they found it is very expensive to build a data system. While it would be their network, most started relying on carriers for a wide area wireless network. Some will build their network and use it for data; this will take a large budget and a lot of planning. The issue would be running the system and maintaining it. That's where the issue lies, in the OpEx. Day to day operations cost money and often drain budgets.

Transportation

In this case ask yourself, "Who are you serving?". That is the real question. I know that transportation has a variety of solutions. I have worked on many deals for many different sections of transportations. So, let's take this into smaller chunks so that we can break down some of the major

Rail or Bus

1) Build your business case for each solution using the final overlapping solution to build the final case.

    a) For transportation, they may need to hire consultants up front to put this all together. It is no easy task and generally, what I have seen is there are many potential solutions. I believe that for something this complicated it may take more than one business case. Different departments could run it or outsourced completely. This business case will need to be thought out, but you could find savings by sharing backhaul or the core by sharing services if you feel it won't cause problems with security or bandwidth constraints.

    b) Build the budget for who you are serving. Passengers want Wi-Fi, open and free and somewhat fast. Operators/drivers want a solid connection to control. Control wants to communicate and monitor the rail/railcar/bus status.

2) Coverage – what are you covering?

    a) Railroads have interesting business cases. They have so many needs for wireless. The good news is that they have quite a bit of infrastructure already built. They have so many needs that are hard to all cover here. Let's cover the major ones. You may want to cover a railway line, a stop, customers on a train or at a station, or a railway yard. I tried to break down the goals so that you have a better understanding of each segment.

    b) Bus routes – here is an option where the bus has a set route. If you have 100 buses, then you cover 100 routes. If you want to offer Wi-Fi on the buses, then you need to have plenty of broadband to the bus itself. If you want communication to the bus, then make sure they have good coverage for the entire route. Usually, a bus would have an emergency call button for emergencies; then you want to make sure that it takes up the least amount of bandwidth but goes the farthest. Maybe the emergency call could connect to multiple systems to send an alert over multiple systems for redundancy. An IOT application that should be available on all buses. It could also be used to track the bus. Each application may have a different coverage.

3) Goal based on each solution:

    a) Customer Internet access – this is where you want to serve the customer, to gain new customers, by making the travel experience a real joy or a place where work can get done. It may be Wi-Fi or LTE-U in the train or bus station or the rail cars or buses. Do you want a very good experience? Many, if not all, railways offer this today on passenger trains so that people see more value in rail travel. In the USA, they offer it using the carrier for backhaul. Railroads rely heavily on the carrier to handle the backhaul. This system is reliable and seems to work. When railways build out customer coverage, they often try to piggyback some of their data on the same backhaul. I get it because it saves money, but it could leave an opening for a security breach. This network should be for the customers then I think it should be dedicated, but it costs money to build and manage two networks, so let them share backlinks.

- b) Railcar business and security – this is where the railways often share services with the internet access backhaul. I said business and security because they want to run all of this over the same network.

    i) Business

    (1) Billing for tickets, food, drink, or anything on board via credit card.

    (2) Updating of times and routes.

    (3) Tracking customers going on and off the train or from car to car.

    (4) Updating time sheets and other tracking functions.

    ii) Security

    (1) Live video feed of each car and the drivers.

    (2) Emergency voice channels.

    (3) Emergency text channels.

    (4) Emergency alerts for drivers and/or passengers.

    (5) Emergency control or stopping of the train if necessary.

- c) Bus video, tracking, and emergency calls could be the most valuable business case for these systems. Each may have different needs and budgets. If you need video for security, then make sure it's connected to an NOC most of the time. While you don't need to upload video real-time, it is a good idea to have it upload on a regular basis. When I wrote this, most bus systems relied on a 2way radio system for their voice, but that could change in the future when 5G systems become built up along the routes. Also, how many drivers already use their personal phones while working? Think about it.

- d) IOT - Monitoring of buses, engines, rails, cars, wheel temperature, or anything that IOT could handle. Railways already monitor this quite a bit. They monitor the heat of the wheels, speed, the rocking of the cars, all sensor-based measuring of how the equipment is operating for critical for safety, maintenance, and accident prevention.

4) Budget – building a budget here goes beyond just building a wireless network. You need to make sure that you have the solution for each service. Voice is going to need a real-time network with little delay whereas your sensor based network will need less bandwidth but perhaps low latency as well. The internet offering will need high bandwidth, but the latency becomes less of an issue. If you share the backhaul can you prioritize the voice and sensor data over the Internet access data? What is your CapEx for the backhaul? Will you build your network outside of the stations or will you rely on another operator or carrier to supply that for you? Will you have spots where there is not coverage and can you live with that? These are all issue to discuss when building the budget. How do you prioritize these services? Make it all part of the overall budget. What I would recommend is

that you break each service out and build your requirements around each around one at a time, then prioritize. It will take a combination of wireless and networking solutions to put it all together. It will also include a way to segregate the networks without causing problems.

5) Spectrum for rail would vary.

   a) In the bus or railcar, they would probably just use Wi-Fi. Maybe LTE-U will be more common then. I have worked on offers where they have two dedicated wireless networks. One for the workers, which is secure, and one for the passengers, which is open Wi-Fi. I don't see this changing too much. LTE could be an option, but again, LTE-U in the ISM spectrum.

   b) For mobile coverage, they may use something in a licensed band, perhaps 3.5GHz unless they strike a deal with a carrier, then it would be one of their bands. It could be set up along a rail line to contact the cars. Some railways use the carriers' LTE to connect because the coverage is great. If they can build a CapEx model around that and get the speeds they need, then why not. The OpEx may be more, but the CapEx should be reasonable.

   c) Inside the train stations is another story, they would probably use Wi-Fi for any customers. For private use, I am not sure. They currently use LMR or a trunking system for voice. I am not sure when this would go away. If it works, they will keep it for if they can. For stats and sensors, they may want something in the 915MHz band, like what ISM offers. SigFox uses this band, but it is a great IOT band and one that could cover a large building with no problem. Lighting control, rail sensors, and door alarms are just a few things they could add to the 5G network. This, along with live video all along the station for security. They could use mmwave if the devices add it. Until then there wouldn't be much point, because who would use it?

   d) To connect the moving railcar, they could use mmwave. In this case, mmwave might be a viable solution if the railway can put it all along the line. It will offer high bandwidth and could connect the cars at high speeds. It just has a very short reach. They may have to put it every mile or so. It could give large amounts of bandwidth to the rail cars making the broadband experience on the cars an amazing experience. A great plan assuming the FCC doesn't ask for billions just to get a license. The auctions have allowed the carrier to get spectrum that small business and industries with smaller budgets could not. I am hoping that the mmwave, which is only good for a limited distance, is available to all, not just the multibillionaires.

Highway

1) Build your business case using the above. It outlines the tools that you need to outline your business case for your solution. Your business case will change as you collect more and more data and align the goal with your budget. Then you need to walk through it again and again until you're sure you have a great solution that fits your budget.

2) Coverage – what do you plan to cover, again, this will come down to the goal. Are you just worried about the tolls booths or are you counting vehicles or maybe you should connect emergency call boxes. These are things needed for the highway. There may be a need connect highway department vehicles or emergency vehicles, although that is usually a separate network. Live video is a huge demand when highways want to monitor traffic conditions or construction sites.

3) Goal - think about what you need to connect and if it's the entire highway or if there are just hotspots.

   a. Counting cars can be done near toll booths or admin offices if that where the need is. You may already have something in place to connect this function, and it is a low data solution.

   b. Emergency call boxes are something that many highways have in remote areas so that drivers have a backup to their cell phones.

   c. Tolls – not all tolls are collected at a toll booth anymore. Many are out on the roads to collect tolls, so cars don't slow down, what happens if someone does not pay and just rolls through? You need with a way to capture video or a license plate, or a way to track that car so that the police can find them or, like in Colorado, tolls can be collected by capturing the license plate and sending the bill directly to the car owners home. By the way, when you do this with a rental car the rental company will send you a bill, I'm telling you this from experience. So, video capture becomes critical for services like these.

   d. A live video is a crucial tool for highway departments. They can use video to monitor live traffic conditions or to track a wanted car or to monitor construction areas. A video is a great tool, but it should be set up properly so that it doesn't use too much bandwidth. Also, think about how you want to distribute the cameras and how fast the video capture needs to be for the speed of the vehicles. Make sure that you considered this when putting together the system. Then the servers to capture the data and whether it will be monitored live by just the highway NOC or also released to the public. While this goes beyond the wireless system, I think it's a good idea to look at the end to end system to build the OpEx and CapEx budgets.

4) Budget – Look at the goals, build the budget based on the solution. If you're covering a long stretch of highway then you have different needs and a larger budget, be it CapEx or OpEx. To break this down look at three things.

   a. Installation and Setup – this is the initial CapEx to install and set it up. The upfront costs are straightforward, and you put these costs in the budget, here is your CapEx.

   b. Monthly reoccurring – here is where you build a monthly budget based on maintenance, backhaul, support, rent, or anything else that you need to pay for specifically for this service either monthly or annually, your OpEx.

   c. Replacements – I am talking about repair but how long until this product needs to be replaced or overhauled. The equipment will not last forever, maybe physically it will, but as for the solution, it may become obsolete sooner than later. Ask this question as soon as you think you want to move forward. I'll give you an example; some highway departments began to use smartphone Bluetooth to track vehicles. While this technology is viable, there was an argument made for the security of devices and how many people will have Bluetooth on at that time. Something that can make a good idea go bad. Some new solutions may have to be put on hold because they need to be vetted or others may be going obsolete soon. Do your research!

5) Spectrum could vary depending on what the purpose is for the highway.

   a. For security cameras, they would probably go with Wi-Fi or LTE-U in the spectrum that doesn't cost anything. It makes sense, and it is what they are doing now. Why spend the money on the spectrum? Mmwave would be an outstanding solution for both fronthaul and backhaul to any video system.

   b. Toll collection will continue as it is for some time, but there are other ways to collect tolls. For instance, some states take a picture of your license plate and then look up the vehicle owner and then sends you a bill for the toll. To take the video, you would need broadband. There are plenty of spectrum options. It could be Wi-Fi or LTE-U. I think mmwave would be a great solution for the shorter links. Maybe FirstNet would allow them to put something like this on their network for security if they could monitor vehicles for security purposes. There is no need to purchase spectrum unless it is 3.5GHz which allows the dedicated spectrum to capture video without interference and at a respectable cost. Just a few options.

   c. Counting cars is a big feature. So far, the best way to do that is video. Let's back to the video option for spectrum. There could be alternative technologies, like motion sensors or weight sensors, but video seems to be something that they can count cars, use for security, and track traffic conditions. It seems like a win all the way around.

County and City Transportation

1) Build your business case based on your need. When addressing the overall business case, remember that most municipalities do not have the resources to maintain a network like this yourself. Remember that you will need to pay for support and maintenance for the wireless equipment or you need to train people to maintain it. Either way, this is the one thing that most teams overlook because they think that it's something you set and forget. You don't; it takes maintenance and care to make sure that the equipment works as well as it did the day it was installed. I would recommend looking at rail and bus above for a solution. If you're looking for emergency responders, go to that section.

2) Coverage is straightforward, what you intend to do with the network and where you need wireless coverage. Follow the goals to decide how to build the coverage. If you truly want to have coverage along with a section for keeping communication between a county transportation but I would think for this section, you would like to track vehicles.

3) Goal – look over what your need it and how you can put it out there.

   a. Traffic light control – often where there is no network connection at the light. You could put in a wireless modem to control the traffic lights and do real-time changes.

   b. Counting cars and monitoring traffic flow are very popular business cases. Cameras are mounted on the lampposts because you have so many resources in place.

Sharing the existing resources like the poles, power, even backhaul. Then decide how you will count the cars, will you use sensors or video? What is the goal, to monitor the flow of traffic and make changes or to just build statistics for future use? Put together what the result will be then you can build a system that will make sense for your solution.

    c. Video feeds – often video can tell you more about the traffic control and to keep an eye on the flow of traffic. It may be the way to count cars and see what is happening real time on the road or to track suspicious cars. It could also be set to allow the public to look at the cameras as a service of the city. One thing that is popular is live video feeds on buses so that they monitor for problems or violence or missing people. It is becoming more and more popular to use these feeds to "see" what is going on and track people.

4) Budgets need to build around the city's budget but not only for the CapEx but the ongoing OpEx as needed. In this case, as I have seen with many cities, townships, and municipalities, they will look at the CapEx for the installation but then expect the equipment to work with minimal maintenance and little supervision. I would recommend to have a maintenance plan and think of expansion, but few smaller cities rarely do that. They lay the responsibility on their IT people along with their maintenance people to maintain the equipment in the field.

5) Spectrum for the government should be straight forward. They often use what is available. Voice applications have licensed bands. They may switch to FirstNet to get better coverage. Data applications may use 5.8GHz to a licensed backhaul and then a specific last mile band like 5.8GHz multipoint or 3.5GHz multipoint. I am not sure if mmwave would be much of a solution unless they need the massive bandwidth and are willing to install and maintain such a system.

Air Travel

1) Build your business case based on the timeline to roll out the sites and install in the planes. Also, be prepared to do updates and upgrades because the technology will continuously improve. Remember that while the passengers on the plane will be the end user, they will complain to the airline if there are any problems, and that will lead back to your group to troubleshoot losing time for ongoing maintenance.

2) Coverage could be on the plane like they do with Wi-Fi. Straight forward and easy to design if the equipment is rated and safe for air travel. The other option that I would like to cover ground to air. More planning and more money are needed to cover air travel.

    a. Ground to air is quite different that covering the ground. The antennas need to face up, but not straight up but maybe at the horizon.

    b. Make sure you know the flight lines. That is the way to look at the tower selection. Looking at the towers along the path should work. Setting up towers on the perimeter will have a better chance to cover more by looking at the horizon. What does that mean, the perimeter? It means that you could shoot the horizon for a larger coverage area. The antennas will be tilted up but not facing too far up like maybe a 10 degree up tilt to cover the overhead and the horizon. With more coverage towards the horizon, you can cover more area.

3) Goal

   a. Understand the routes. Is their goal to have the maximum bandwidth, low latency, or reliable coverage? The coverage is usually the key. Chances are they want all 3, but they will prioritize, or you will do it for them.

   b. Understand that the latency will be higher and that the vehicle travel speeds are high. When choosing equipment, the speed of the plane and the distance will all be factors. Also, think about how the equipment gets mounted on the plane. The plane will need to have FAA-approved equipment and antennas that can be mounted low profile to a plane but also able to have coverage to the ground at high speeds.

   c. The desired bandwidth may vary on flights by the time of day and routes, so take that into consideration when planning the system.

   d. One thing that I don't see is any live video feeds on planes. That is something that could help in a time of crisis. It is something that could be added to planes so that remotely they could turn on video feeds.

4) The budget will be more than building out the sites on the ground. The plan will include what will be on the plane. Look at the end to end system. Your plan may be to have the two systems completely separated, which is fine if someone is looking at it end to end. Obviously, these are two very different processes, but they should be looked at together when doing coverage and building your budget. The planes need to be worked on while on the ground and scheduled around the airline's maintenance schedule whereas the ground work can roll out when the leases and site acquisition is complete. You should know by now that site acquisition takes a long time, but a project like this will take a very long time because the airlines are hard to break into and get equipment approvals. Prepare for a long lead time on a project like this. Also, prepare for ongoing maintenance and support and upgrades. It's not a system that you install and forget it. You will be supporting people that have problems; you will continuously increase speeds and improve equipment as well as troubleshoot problems. You have all the problems of the site issues, like lightning, and on the plane, like equipment failures. This system could be stressed quite often.

5) Spectrum for airlines is tricky. While they commonly have Wi-Fi inside planes now, the ground to air may be regulated by the FCC and approved by the FAA. I would like to see 3.5GHz used for this, maybe mmwave would work, although it may be too far. I don't know what guidance to give here. I would imagine 3.5Ghz won't be used due to the nature of possible interference, then a licensed band with large amounts of spectrum should be used. Generally, from ground to air they used licensed spectrum. It is the norm for air travel because the FAA must sign off on it.

The 5G Deployment Plan Handbook

Unmanned Vehicles

Drones/plane

1) Build your business case around what the vehicle will do. The Wi-Fi controls most drones that you buy off Amazon on the phone. They did try Bluetooth, but the distance was limited.
    a. In this instance, we are going to talk about a drone delivery system. There are many options here. The way I see it is that the drones will be given a flight path ahead of time. The drone needs to be preprogrammed with the delivery address and a flight path. They should not go line of site, but flight paths will need to be defined. The NOC could communicate with the drone to monitor progress and for feedback, mostly in the early days. Drones will need to have a way to talk to each other, maybe with Bluetooth but there is a need for instant connectivity or a radar system so that they don't run into each other. They will also need to sense weather; rain could be a huge problem. I don't see the need for video over wireless, but they may keep it onboard to capture proof of delivery, and for security, in case someone tries to steal the drone or package. While video sounds great, the reality is that to handle all that video would be a wireless traffic nightmare. For one drone, it's great, for 100 drones, it's a nightmare. All they could do is put a recorder on the drone. Now, it is not efficient to send video from a hundred drones back to the NOC. I could see a failsafe system built in and I see the drone only capturing video of the delivery.
    b. Emergency responders need drones for so many reasons. The police could use them for pursuits but the real need for drones, from my perspective, will be the search for victims in a disaster. While the Wi-Fi coverage would be fine for a one on one coverage situation but there will be a need for wide area coverage sending back video. There is a need for more than normal video but also UV, heat sensors, anything that could detect a human life in a building or rubble. It would also require drones to take flight in nasty conditions, to go where humans can't go or won't go. Safety matters. I also see a fleet of drones, not just one. We can't limit ourselves because we're saving lives. So, it will take coordination and planning but think if they would concentrate search efforts for survivors to be where survivors are. Not all over or waiting for a search dog to detect someone. More territory in less time and concentrate the rescue party where life is detected.
2) Coverage is going to be determined by the need. For delivery systems, it will be the coverage area for the delivery service and chances are it should be close enough that the drone can come back and recharge. Battery life will play a part. If it's for emergency responders than the coverage will be very specific, a city, a region, or a highway. One thing I would recommend is that you understand the coverage for the need and test it.
3) Goals for this system are very specific. They think through this but don't make it complicated. We are using the drones for delivery, period. We are using the drones for search and rescue, period. It should be straight forward and simple.
4) The budget covers what the end to end system will cover. Will it be just using Wi-Fi or LTE from a carrier or a private wireless network? Obvious, a private network would be very expensive but would allow you to control everything. This budget should be thought out like this if it's your network you will have heavy CapEx then OpEx to continue to operate and

maintain. If you use a carrier, then the CapEx is very low, but the OpEx could be very high. If working with a carrier using LTE or 5G spectrum make sure you test the coverage. Make sure it's what they say it is.

5) Spectrum will be determined by what the goal of the business case.
    a. Delivery always sounds great for Amazon, but so far, I have yet to see anyone like that, Amazon or Google or anyone outside of the carriers invests in spectrum. When the US government asks for billions up front, the customer list gets very small. The FCC found a way to keep the smaller carrier out and limit who can afford to build a system. It appears that this philosophy has helped the US pay down the debt but squash fairness in wireless. Sorry, I digress. I see the carriers supporting this type of network using their spectrum and charging a monthly fee. Heavy on the OpEx and light on the CapEx.
    b. For emergency responders, it would be a different story. They could use a carrier or a FirstNet system if they survive the disaster and cell sites don't overload. They would need to be sure. In the case of a post-hurricane or earthquake system, they would probably need to set up their network. Could be Wi-Fi or 3.5GHz or anything that they have access. In this situation, it wouldn't matter because it is an emergency. However, it is something they should plan (ahead of time) and have immediate access to the equipment. One thing that I have seen with emergency responders in the past, they tend to invest in a nice system then put it in storage for a long time and expect it to work when they pull it out. Unfortunately, it needs to tested regularly, twice a year, updated and verified that it still fills the need. Like so many departments learned, updates and testing are critical. I see them do it with their equipment but let the wireless and computer gear lapse into oblivion.

Automobiles

1) Build your business case around what the vehicle needs. Here in the USA, the Department of Transportation regulates much of this for the automobile industry.
    a. Autonomous driving is a new ball game. I would not trust any carrier's system to do this without testing. However, the driving would be mostly about the GPS system and updated maps in the vehicle. The wireless would have more to do with autos communicating to each other. You would also need to have it connect to a carrier's network to be sure that it is on track, safe, and get any updates that are available. Google did a great job with their self-driving cars, but you still need to make sure the vehicle is connected.
    b. Car emergency connections will be like what they are now. AT&T has the OnStar backhaul now, and it works great. They have a good track record. There are others out there with a similar model. While it can help with directions and other things, the big thing it offers is an emergency response and accident detection. If you're in an accident and the system still works, it will call for help before you can and maybe before anyone finds you. That is a great feature that will save lives.
    c. Internet access to the car. We all see this with Wi-Fi in the vehicle, for passengers, but it could also feed the entertainment (music) or the GPS with updates. It could also send alerts from the vehicle to the dealer to let them know about any engine problems. If they see problems in a massive number of vehicles, they could issue a proactive recall

and let the owner know that they need to come in for service via email or text. While I think it's awesome, I only see the carriers offering this setup.
   d. Maybe an easy play for people outside of the carrier or auto manufacturer realm would be to create apps for the vehicle. It's a thought. Maybe you could create a device for the vehicle.
2) Coverage for the vehicle would have to be everywhere. Too hard to limit this. However, the cars should be able to talk to each other. I understand that they will have video and radar to sense the surroundings. If they could all talk to each other than the car that is 100 feet ahead could tell all the ones behind it there is a problem or accident or conditions are clear. That is going to be a mesh system of some kind.
3) The goal of this offer is more about what the solution will be.
   a. For autonomous driving, it is more about the connection to the cars for updates, but also about the cars being "aware" of what is around them. They may need to talk to the car beside them to see where they are and what is happening ahead of them. Do they need to apply the brakes, is there an accident ahead? What will they need to do?
   b. Car emergency connections will need to be considered to peer to peer connection will be very important. This spectrum needs to be clean and short.
   c. Internet access for the riders will need to be on a wide area system. I would say the carriers have the best chance at connecting this service.
4) The budget for the wireless systems will need to be thought out as an OpEx offer. Most vehicles are going to rely on a national carrier for most of their needs. The peer to peer will probably be license free and built in, like the Bluetooth on your phone. The automakers will have devices installed and the vehicles. What you need to worry about is the OpEx of the carrier's fees and what service you can provide.
5) Spectrum would be whatever the carriers offer.
   a. Wide area coverage would be a carrier. If you were to offer something, it would need to be a spectrum that could cover a national footprint. If you go licensed, expect to spend billions unless the FCC (in the US) starts giving small business a chance. It could be possible for narrowband LTE spectrum. That would work fine for the control and low latency that autonomous would need to communicate as close to real time communications as possible.
   b. For internet access to the vehicle, it would also be wide area coverage. Chances are it would be licensed LTE carrier that the nationwide carriers already have. If you were to try to cover it with Wi-Fi, it would be too hard to build out in my opinion.
   c. For peer to peer, it could be 5.9GHz which is being approved by TIA and accepted by the department of transportation in the USA. That means that most vehicle manufacturers are going to have this in their vehicles. To learn more, https://www.standards.its.dot.gov/Content/documents/workshop_CVRIA_pres_06-10-15.pdf and see what they have to say. It is a peer to peer technology called Connected Vehicle Reference Implementation Architecture (CVRIA). Learn more at https://www.standards.its.dot.gov/content/documents/V2x_standardization_plan.pdf if you're interested. Short range, peer to peer, wireless technology.

Boats

1) What will the business case cover?
    a. Controlling boats has been done, so it's not new. There has already been a business case built for boats. In canals and lakes, there have been boats controlled using the ISM band and Wi-Fi. It had to be a line of site, (LOS), but it has worked well. In every case, they kept someone on board in case something went wrong or if the weather got bad.
    b. Offering Wi-Fi on boats is common in the US. In the Gulf of Mexico, they use the oil rigs to pass the LTE off to boats. It's being done across the world and is more common that you think. This service can compete with satellite because it's more cost effective and now with the higher speeds it can look quite attractive. I think that ships will need satellite but when they come in range of the signal they can get higher throughput. For the smaller boats that are in a body of water, like a lake or a gulf, they can use this for internet access and cell phone service. The company that I know about is Tampnet, http://www.tampnet.com/gulf-of-mexico/ and they offer service in the Gulf of Mexico and other areas of the world. INET also did a trial, http://www.prnewswire.com/news-releases/infrastructure-networks-inc-announces-pilot-lte-deployment-in-the-gulf-of-mexico-300043579.html to see how it would work.
2) Coverage for this system could be from shore to water or in the case of the Gulf of Mexico; oil rigs are the hotspots. Backhaul on the water could be satellite or point to point. Point to point microwave does work from oil rig to oil rig.
3) Goals need to defined. If you are offer lake boats Wi-Fi from the shore, make sure you have the coverage and think of the latency that they might encounter. They may use something like Skype to communicate so latency could be an issue for voice. Don't make promised you couldn't keep. If you offer LTE for phone service, the same applies. Offering internet access will be different than offer voice service. You also reach matters, where will you be
4) Budgets are determined by the service you offer.
    a. For internet access using LTE, you will need to think about the spectrum you have and what is available. Why does this matter? Because having the core allows you to offer multiple services. Your CapEx will be whatever it is to deploy in the shore. If you are deploying on oil rigs, then you need to have people trained and certified to work on oil rigs. If you plan to lease spectrum off a carrier then your OpEx will increase, they will get paid for usage, so you will need to work out a deal with a carrier in the US or try to win spectrum in an auction.
    b. Is it shore to ship? If mounting on shore, then it would be straightforward for most of you. Just take into consideration the dampness of the site, and if there is salt, the salt will eat most metal. Since your OpEx will rise with the increased maintenance costs. Just things to think about as you move ahead.
    c. 5G spectrum may allow you to do more, like mmwave if that is going to work. What will it cost to make that work? We don't know at this point.
5) Spectrum is a funny thing.
    a. For LTE access on a body of water, you still need a license. You could lease spectrum from a carrier on water. The way the auction in the US worked most carriers got spectrum over a body of water. Maybe you could to make a deal with a carrier for spectrum that they are not going to use. I believe that Tampnet considered using

spectrum from a major carrier in the US, like Sprint or AT&T, to cover the Gulf of Mexico.
b. I have seen WISPs cover lakes and canals with Wi-Fi in the ISM band, specifically the 5.8GHz and the 2.4GHz spectrum. Working for a short distance and could solve a solution with unlicensed spectrum. It is a viable solution that is cost effective and quick to deploy. When LTE-U gets going, it may be a viable option that would allow more useable bandwidth.

Emergency Responders

1) Separate what it is you want versus what you need! Business cases will need to be laid out for approvals and budget building. What will you use it for, voice or data?

    a. Voice: To be clear, first responders need a mission critical system for voice communications. It would be very hard to make a mission-critical 5G system, or even system mission-critical LTE, in 2017. Maybe I should say it would be too expensive now. FirstNet didn't do it, and they had no intention of doing it out of the gate due to the high cost. LTE mission-critical was too much money. So, if you need mission critical, it is a lot of money, so this may not be the best system for voice communications.

    b. Data: Build your business case based on data you need and budget. Being in the public safety sector makes your budget tight. While you need critical reliability or someone could die. The business case is critical, you to get your funding. The design determines the critical communication acceptance. Therefore, so many states in the USA are looking at building their system. The FirstNet model takes control out of the hands of the people using it; they may feel like they have the same coverage with a commercial carrier. However, your municipality could put in its 5G system if the spectrum is free or license free. If you think public safety wouldn't do this, they do. Wi-Fi and 4.9GHz systems were deployed by public safety entities all over the US because it's cheap, easy, and a great way to connect laptops and data. Look around, budgets matter and sometimes you need to get creative. Video on a street that trouble happens helps. I once helped a group put a video over a bar because they had a lot of trouble there. The police didn't think the bar owner would go for it, so then they just asked. He was more than happy to put it in his second story window to prevent trouble. He wanted the trouble to stop, and it did. The police could monitor the camera when they were somewhere else, and they did, and show up when something happened, and they would. Great system and solution all done on Wi-Fi.

2) Coverage

    a. Each state or municipality or even FirstNet must figure out what coverage is needed and wanted. Where do they cover, who are they protecting? What coverage do they need?

b. Indoors you need to be sure that you cover the buildings you need. Indoor coverage is often an afterthought of the system planning. The budgets are tight when they start, so most people think of outdoor coverage. Often, in public safety, they must build within their budget. They can't design a system then ask for money. If they design first, then they shrink it down to fit the budget.

c. Think of the public, do they want you to monitor a troubled area, even temporarily? You could deploy cameras, but you need broadband to capture it effectively. Maybe Wi-Fi, but you may need something secure and licensed. Maybe FirstNet LTE or 4.9GHz or maybe mmwave. All the same, someone should record it, so you need the backend services to support it.

3) Goal – decision time:

   a. Decide who you are helping. Police and fire have similar needs but maybe different coverage areas. Paramedics have a different need and need different coverage, maybe only in the ambulance compared to remote at the patient's location. Fire rescue and police need serious building coverage as well as mobile coverage. For these services, it could mean the difference between life and death.

   b. Decide where you have the largest need. By this, it could be coverage or a specific department. It would be for a specific application.

   c. Decide what service needs the most help. Police, fire, paramedics? Who needs it more? Who will have priority?

   d. Decide if you can share the services among first responders. For those of you outside of public safety, this may seem strange, but the services each usually have their communication service. It has been that way for a long time. Police have their system, fire departments have their system, and paramedics may have their system. Each department has a budget, and they don't always share. LTE is supposed to change all of that because it's a system that all emergency responders can access if the spectrum's dedicated to public safety.

4) Budget is going to vary on how you proceed. In the USA has an entity called FirstNet that is building the nationwide LTE system. They have most of the control of the nationwide network. It affects your budget because public safety entities may have little control over FirstNet's coverage, but their CapEx could be very low. CapEx would be laid out for devices and setup. The OpEx, on the other hand, may be very high. Now you must pay monthly bills for service and coverage. Think about who needs it. Also, if a USA group goes with FirstNet, make sure the coverage is where you want it. The budget may look good in the beginning, but if the coverage sucks, then you're stuck paying for crappy devices when you could have gone with a large carrier for like service.

5) Spectrum is a tough call for first responders. For emergencies, they still rely on their voice for most activities. The data's used for research and support. Paramedics would rely on these systems for patient information. Police need to see pictures. Fire and rescue teams

may need to see blueprints of a building. All of them need quick access to records which makes a case for broadband.

   a. They could use Wi-Fi, LTE-U, or the 4.9GHz band to capture video. That is obvious which is why they are doing it today. However, with the improvements of the LTE from the carriers, they rely on it less and less because the carriers licensed LTE works so well and are becoming more and more affordable.

   b. There is the 700MHz that FirstNet has. Currently expected to be 4G LTE but soon they will be 5G ready. If you go with FirstNet, you need to consider a few things. They may not be a true mission-critical LTE system. It is more like a carrier. Does it make sense to go with FirstNet or another carrier? That's going to be a decision you should make, and chances are it will be a financial one considering your OpEx budget. Will FirstNet be more cost effective? Will it replace the carrier's system? FYI – the FirstNet system will be a carrier's system until the 700MHz system gets built. Also, it does not meet the requirements of mission-critical LTE, not yet anyway. Maybe in the future, it will get there, but it is too expensive to build a nationwide LTE system. Maybe if the states could have done it then, it would be mission critical.

   c. The narrowband LMR systems are working well for voice, but they don't fit any needs of 5G now. Maybe they can be repurposed for IOT in the future, but the first responders are going to rely on LMR for some time.

WISP

1) Build your business case based on the criteria set above. Decide your future for expansion and growth. Plan it out on paper.

   a. Budget is a big concern here. Most WISPs (that I worked with) run on a shoe string budget. They often use Wi-Fi because the equipment is cost effective and the spectrum is free. So why not?

   b. Plan for small coverage cells and decide who you will cover and what they will pay. Build a solid plan around what you know.,

2) Coverage will be determined by who you intend to cover. You build this system on customers that are already asking for this service. You may have a select group of customers or buildings where you intend to cover. You may sign agreements to roam onto other WISP systems to make the customer think that coverage is larger than it is. It takes a lot to build out a large system. Just look at what Comcast has done, that is amazing.

3) Goals for WISPs will usually have different goals. If they want to cover a specific area, they usually target local customer in that area for home and maybe for Wi-Fi access on their smartphones. If the WISP gets LTE-U, then calls would have a better handoff from the carrier to the WISP. While this is not an ideal situation for the carrier, the WISP could try to get a roaming agreement with a carrier. This trend is starting to slow down as carriers build out better and better coverage for their LTE systems. They no longer need many WISPs, just ask Comcast. While Comcast is not a WISP, they do have a very large Wi-Fi network that

most carriers have shown little interest. So, when putting together a business case as a WISP, decide who locally you will be serving. If you're in an airport, then it's the travelers. At a bus station, travelers. In an apartment building, it's the people who live there. If it's a rural community, then it's the residents, probably at their homes or in a diner or coffee shop. Just think it through and remember that coverage matters. The main goal should be coverage for local customers.

4) Budget building for WISPs revolves around the customers, what do they need? Let's review some questions below.

   a. What can you offer?

   b. What is their primary application?

   c. Then look at the assets, like towers, monopoles, and building tops that you want to mount on. What is the rent? Are there zoning and permitting issues? Will you have 24/7 access if there is a problem?

   d. Who will maintain the site? Do you have a company or will it be one of your employees?

   e. Do you want to expand? Will the sites you selected serve your current customers?

   f. Do you have a backhaul plan?

   g. How will you monitor their traffic?

   h. How will you bill your customers?

5) Spectrum for WISPs is Wi-Fi. Maybe they will adopt LTE-U or 3.5GHHz now that there is more spectrum available. I am not sure, but Wi-Fi is the standard. It's available, free, and the equipment is common off the shelf, (COTS) equipment that most any company could afford. It's the sites that may be a challenge. It would be great if you could get something dedicated, licensed, and with more power than Wi-Fi, but you will pay for it. We need to look at all options before you roll out because you may be missing an opportunity or a way to differentiate yourself from others. If you can get licensed spectrum, even lightly licensed, it makes you look more professional than the guy using Wi-Fi for everything. Also, if you plan to use wireless backhaul, then think about that spectrum as well. If you plan to use 5.8GHz ISM band for everything, then you need to plan the channel usage ahead of time.

Small Carrier

1) Build your business case based on the criteria set above. Decide your future for expansion and growth. Plan it out on paper. Chances are if you plan to build for mobile coverage you will need investors. If you planned to be a fixed wireless carrier, then you will have a customer in mind before you build, or at least you should.

   a. Mobile or fixed coverage is the big differentiator. Low latency or medium latency, does it matter for your system. Then build a plan around those factors. That will help you decide the business case.

b. Will you purchase spectrum, rent spectrum, or are you going license free? It will cost you more money to go licensed, but it will make you a real carrier and not a WISP.

c. Will you build out LTE everywhere to start? If so, you need to think about backhaul, the core, and maintenance with upgrades.

d. Build a timeline of the rollout and release dates. Can you become an MVNO and ride on someone else's network until you build your network? Yes! It has been done in the past and could work for you.

2) Coverage will be the area you intend to cover, obviously. If this is for mobile coverage, then what area or region will you cover? You may have a select group of customers or a region or even specific buildings where you intend to cover. Your income stream will decide. Why cover an area with no customers? If you are a small carrier, then think about the T-Mobile model, they cover the urban areas and roam to the rural areas on another carrier because it makes good business sense.

a. Roaming will be an issue. Will you have an agreement in place with a larger carrier? What is your plan for roaming?

3) Small Carriers will need to cover an area where they have licensed some spectrum or leased it from another carrier. Usually, they also have some roaming agreement with a major carrier. The smaller carriers can offer only so much coverage. They build out their area well and then if someone leaves the area they roam on to one of the big four carriers. It goes both ways. The big four will also roam onto the local carrier for better coverage in a region, city or area. The local carrier makes up for lost revenue. The smaller carrier must think through their strategy. At one time, they thought that outdoor coverage was sufficient. With the coming of 5G, the indoor coverage for high data usage will be a game changer. Suddenly when a national carrier sells to a company, they may need indoor coverage and the person with the building coverage will have the best coverage. Seriously, think about what you should gain. If the carriers don't work for you, then maybe you need to find someone who will. The main goal should be coverage for the region.

4) Budget building for small carriers will be much more than building the system out. It will be for spectrum, maintenance, roaming fees, all the things it will take to maintain the system and customers. Look at the equipment, sites, and backhaul to determine the CapEx. I will tell you that to build a system for mobile users is a lot of money up front. It takes time to plan and should have investors fully informed of your 5-year plan. They will expect to see how you plan to expand. You don't necessarily have to expand in coverage but in service offerings. Also, don't think that you need to provide all mobile coverage. You could become a fixed broadband carrier to complete with cable companies for in-home internet access. It would be a great alternative to offer someone wireless connectivity to the home.

5) Spectrum for Small Carriers will be licensed spectrum. They use the spectrum they own or rent. They could sign a partnership with one of the carriers to use their spectrum in a specific region. I believe that Verizon and Sprint have partners that use their spectrum. This way the smaller carrier can build out the system for the larger carrier and make some

money on it. Don't limit yourself, though, think about what you could do with new broadband spectrum. If you could get mmwave, then you could provide some options for fixed broadband to the home. You could compete with cable and FTTH, (Fiber to the home), by providing either LTE TTH or 5G TTH and become an internet provider to residential or business customers. Think about new avenues of income that new spectrum could make available to your company.

IOT Systems

1) Budget building for something so new will be a challenge. IOT is a very exciting industry and one that is already rolling out. SigFox has already rolled out an outdoor IOT system across Europe and the USA. It is a viable business plan to offer service to machines in a low latency low bandwidth offering. Spectrum could be license free or low-cost narrowband spectrum. Too many options make budget building complicated. So, look over this list of questions.

    a. What equipment will you attempt to cover? More importantly, what is the application? Specifically, who is it for and what type of monitoring will you do? The equipment may matter here.

    b. Who is the customer? Utilities may have different needs than health companies or automobiles. Maybe each section has a specific need based on the applications.

    c. Can you use license free spectrum or do you need to purchase spectrum? Spectrum could eat into your budget.

    d. Is it indoor or outdoor? Indoor is straight forward and would be a case by case budget. Outdoor coverage is another world because you need to have a target demographic of who you want to cover. Huge budget concerns outside for OpEx and CapEx.

    e. What type of backhaul do you expect to have? If it is low latency, then speed matters so that the cloud will play a big part.

    f. The core and the cloud will be a part of the expense. Running a backend with the apps and programs needed to support the equipment, monitoring, and customer alerts will all be part of the budget for starting up and expanding, upgrading, and maintaining. Look at the big picture beyond the build.

2) Coverage is going to be determined by where the expected customers are. We could look at the IOT deployment as serving customers that you already have. You would not just build it hoping they would come, but you would have a target customer base and build around their needs. Then you can try to add more customer to the system. This system is going to be pretty "customer specific." If you're doing indoor coverage, then it's obvious but outdoors is going to take planning around your spectrum, sites, and end devices you need to hit.

3) Goals for IOT systems are very specific. Normally you would build it for specifically for the purpose that the end equipment would need. Do you control the end equipment? Then be sure you have remote control. If it's monitoring, you may tell it to listen or wake at a specific time to "ask" it what the status is.

a. IOT is a term that is used with 5G a lot, but the only thing that 5G must offer IOT is low latency. There are many IOT systems, like SCADA, which are already out there. We just call it "machine to machine," (M2M). Most IOT equipment is low bandwidth and often only talks a few times a day or less to conserve battery life. Narrow bandwidth spectrum is available and very cost effective.

b. IOT is usually low latency because it wants to talk and get a response very quickly so that the device can go back to sleep if it's relying on batteries for power. Why does this matter? Because the distance from the device may play a part in the latency. In the early days of multipoint systems in the ISM band, they would lose connectivity to far buildings because of low timeout functions in the network. If they were too far away, there would be latency issues. Know what the latency will be for the equipment, then take all factors into consideration, not just the equipment, but the distance of the backhaul and fronthaul and last mile. It all adds up, if it's too high then it adds up to trouble. Once again you need to know where the customer's equipment and what latency they can tolerate. It sounds like a lot of work up front, but this is something that may come back to haunt you again and again if you plan poorly. The main goal should be low latency.

4) Budget building for these systems varies widely. If you have your goals laid out ahead of time, then you should be able to look at equipment, sites, and backhaul. I will tell you that to build a system for mobile users is a lot of money up front. It takes time to build a plan that you can show investors. For IOT and WISPs, it usually can be built out based on where the customers are, small and incremental construction is the norm. You could build on need more than coverage.

   a. What equipment will you need for the sites?

   b. What will your site rental be? Don't forget permitting and zoning.

   c. Backhaul and fronthaul may add costs for CapEx and OpEx.

   d. The core and the applications and the monitoring and alerting the customer with problems.

   e. What kind of customer support will you need?

5) Spectrum for IOT is interesting since you have so many options.

   a. SIGFOX and LoRa use narrowband spectrum to talk to the devices out there. Narrowband is generally in the sub 6GHz range. Many current systems use the ISM band 915MHz, here in the USA. It is easy and quick to deploy and works relatively well. Low latency and low bandwidth. Long battery life. SIGFOX is a service provider who is rolling out the IOT wireless system in Europe and the US. Both systems could use lightly licensed bands, but they use the unlicensed bands since it's cheap and easy.

b. Several companies use 2400MHz and 5800MHz ISM because of its availability in the US, it's free, and it carries well. That spectrum's overcrowded with users on Wi-Fi but usable.

c. Wi-Fi is a great alternative for indoor coverage, although it will take time for the devices to catch up. The 3GPP group just release Wi-Fi for IOT which increases battery life. If you need a broadband solution in the free spectrum, this may be the best solution.

d. NB-IOT is Low Power IOT, which means it uses very low power to communicate which enables much longer battery life. High spectrum demands more battery usage. Battery conservation requires narrowband spectrum. The format could be LTE, or it could be GSM. Wi-Fi has a long battery life option. You have choices of formats and licensed bands, usually only 180KHz of bandwidth. Not the massive bandwidth that the FCC auctioned off for billions but rather the less expensive spectrum that someone out there is holding onto in the LTE and GSM bands. It would be 3GPP approved. Let the smaller companies take advantage of this before a billion-dollar price tag gets attached to it.

e. Bluetooth is working on an IOT option since they already connect to devices, but it would be very limited on distance and battery life would not be optimal. It would be an indoor solution, maybe for home use. I am not sure how this would work in an industrial environment.

f. For more LTE, there is CAT-M which the carriers will use. It is a broadband solution just like what the carriers are using now. While 1Mbps would be sufficient, it would be something the carriers could roll out quickly.

g. There is another solution in the GSM licensed bands called EC-GPRS which is a GSM format. Licensed spectrum with about 200Khz of bandwidth. While it is part of the 3GPP solution, I don't know much about it. It seems very few companies went this route, but with LTE taking over it might be a viable solution in spectrum released by the carriers on their way to broadband LTE. Maybe spectrum will start to open, and this will be a viable solution.

h. One more note, think LPWAN, Low Power Wide-Area Network, could be any of the narrowband technologies above, but sometimes people are looking for something different. I look at LPWAN as an NB IOT solution.

Enterprise

1) Build your business case around the needs in your goals. Devices, equipment, backhaul. Break it into CapEx and OpEx for the equipment and the backhaul and the support. What is the purpose? To provide connectivity to the workers or visitors in the office or to track equipment or to control devices. Planning of time will save you grief later. If you plan to add a new service, who or what is the end user? Then evaluate what you have and decide how to move forward.

2) Coverage is limited to your offices. You may have a specific area to cover, or you may have multiple offices that need coverage. I would say it is indoor coverage, but if you have a campus or a chain of buildings then think of how to cover the outside and how to roam from building to building. I am sure you won't have a flat network across all buildings. Also, think about roaming from your system to the carrier. Remember that you control what carrier can come onto your system. They all want your business, but you certainly don't need them. Cover what you need to cover.

3) Goals should be set by the applications you intend to run. If you only want your teams to have network coverage, Wi-Fi or LTE-U may be good enough. What will be on the system? Laptops and smartphones would be the obvious devices. What if you want to track items in the business? Then IOT applications should be added. Do they need batteries to run? What type of object will you track, do they need to be tracked real time? What about video? Video requires a broadband network.

4) Budgets are built around more than coverage. For instance, they are built around devices and backhaul. System budgets may include a long list. CapEx versus OpEx again. Since your needs vary, planning takes priority.

    a) How much wiring will be needed? Mostly for setup, but do you want to think about expansion?

    b) What power do you need? Can you do Power over Ethernet, PoE? Trust me; it's better to use PoE for so many reasons.

    c) What will the backhaul cost to run? To set up initially? What's the design and setup? What routers do you need for this type of backhaul? What are the monthly fees (OpEx)? What about growth one year from now?

    d) What about devices? You may need to add your own devices, and they should have the wireless chips in to provide access using the format your systems use. Will you be adding devices in a year? Three years?

    e) How will you roam? If it is a smartphone, maybe a carrier but a laptop you may decide that any Wi-Fi or LTE-U hotspot will work as long as you have a secure VPN. It's all good.

    f) What support will you need? Can your internal IT team handle it or do you need to hire a company to do maintenance and support?

5) Spectrum for Indoor usage by default is Wi-Fi. With 5G spectrum rolling out you have more options! You could get something broadband that is lightly licensed to be used on your devices or for a specific need for IOT.

    a) Wi-Fi is the old standby and has worked for years. If you're happy with it and it serves the purpose, then stick with it. If you have it, keep it. Why lose it unless you are going to make an IOT application, then you may want to add something new.

b) LTE-U is also in the Wi-Fi spectrum. It may mess with your Wi-Fi, but it may work better with your smartphone. It's up to you to see if it will be worth the investment. If you go LTE-U then maybe a carrier will roam onto your network, and you could have seamless smartphone coverage, if you need it, from inside your campus to the real world. Think about the benefits before you invest.

c) CBRS is the 3.5GHz lightly licensed spectrum. It is a viable indoor solution, and now that the FCC is opening more spectrum, you may want to consider it for your building. Again, this is something that the carriers could roam to in the foreseeable future. I would think you could add it to your building or campus and it might be cleaner than Wi-Fi, but now, January of 2017, no smartphone has this available. So, you would need to think about what devices have this and what doesn't.

d) With the growth of VR, virtual reality, and AI, artificial intelligence, you may need massive dedicated spectrum for these purposes. Here is an ideal application for mmwave spectrum. It should be lightly licensed and would be a viable solution that offers massive spectrum. Putting this in your building would allow you to have spectrum dedicated for special purposes, like VR. If you have this application, then consider adding this to your spectrum arsenal.

Business or Building Owner

1) Build your business case around the goals and budget you set. Also, think about the payback. The process for this is thinking about how to serve your customers. Will they be happy, is this a true value add, will they appreciate the internet access you provided? If your employees are using this will they save time with this? Is this something that people truly see as a benefit? Will emergency services be helped in any way? Put the benefits in writing, so everyone will see more than the expenses, but also the upside.

2) Coverage is going to be covering the building or business. Indoor coverage means you have a limited coverage area. You don't need to worry about outdoor coverage. All you need to worry about it installing indoor hotspots, with unlicensed or licensed spectrum. Also, the carriers may want to roam to your spectrum, so make sure you control it.

3) Goal – the goal here is to provide coverage for your workers, your lessors, or your customers. The key is to provide them with sufficient coverage in your building or business. Think about what devices they will use. How will they be best served? Will they need to use smartphones? Will you have more than Wi-Fi or LTE-U? Will you allow carriers to roam onto this network?

4) Budgets are built not only around the installation but the ongoing monthly fees. Did your account for the monthly backhaul charges? Do you have technical people to maintain the system and react if there is a problem or do you prefer to outsource that function? Do you have a system set up to take complaints and open a trouble ticket? Do you want to deal with that part? Think this through then build the budget on more than just the hardware and installation. Build the OpEx budget at the same time. They should go hand in hand with the planning process.

5) Spectrum for a building owner is straight forward.

    - If you want to offer the residents and visitors Internet access, you will use Wi-Fi or maybe LTE-U.

    - If you want to monitor the devices in the building, look at IOT. There are several things you could do, track devices or monitor meters and thermostats and applications like that. In that case, Wi-Fi may kill the batteries so you may want to look at some of the 915MHz NB IOT solutions.

    - Other options are to work with the carriers to put small cells in your building. Carriers don't pay for this like they used to, so the best option might be to look for an alternative, like the CBRS or LTE-U, something that might be cost effective and yet provide carrier roaming coverage for LTE inside of your building. DAS is still a viable option, but with the world going digital you will need small cells or a digital DAS system to cover properly. You may still need DAS for your public safety system for fire rescue and emergency responders. In many areas, they still use analog.

    - Just to go over it again, remember that emergency workers expect coverage in buildings, which means a public safety DAS system needs to be there. Emergency responders need to communicate in the event of an emergency.

Building Maintenance

1) Build your business case around more than just the initial build. Think about how it will serve you for years to come. I believe that the CapEx will be the thing most people focus on CapEx. However, the OpEx for the future services and the growth should all. Once you have a view of the big picture, then you can look at the end to end solution to show the value and the expense. Value is the key, the value to save money in the future as well as improving service.

2) Coverage is straight forward. You want to cover the building, but where? I see so many companies that cover the building for the wrong solution. They think they know where to cover. I worked with a hospital where the wireless company didn't listen to the maintenance department which caused a huge area not to be covered properly in the basement and the penthouse (elevator room). The wireless company covered what they thought should be covered based on what the staff told them. Meanwhile, the valves and gauges in the penthouse, which is the top of the elevator room, and the maintenance offices had no coverage. It's a huge oversite that had no easy solution. They had to run cables up there to connect the radios. If they had just taken the time to talk to the supervisor, they could have avoided extra costs, a second visit, and night work. Make sure that when you are walking the job and bidding, you talk to the right people up front, ask for all the stakeholders and what is the purpose of the coverage. Who are they covering? Try to talk to the potential end user.

3) Goal – this is to make sure you not only cover what you need to but also to provide the correct service. What you should look understand is the service. I'll dive into this more in the budget but what is the service you will provide? Is it to control equipment, monitor

equipment, only look for alarms or video? Is it for just maintenance or are you going to share the service with other building services? What kind of bandwidth will be required in the building and out to the internet? Is the monitoring going to take place in this building or at a remote location? (Like a remote NOC.) Think what the result will be. If you are just doing the installation, then you should be concerned about power and data being run to the unit, no matter if it's a camera, sensor, or controller.

4) Budgets are based on size and coverage and service. What's your focus?

   a) For one, the equipment, this is a simple count of the product but also what the product will do and how much cabling is needed.

   b) What is the product? Is it going to be video, sensors, or controllers?

   c) The installation will take the most time to understand. Where you need to go and what you need to run. Remember that there is roughly 100 meters or 128 feet, to running CAT5 data. It may be shorter if you're running PoE. On the other hand, if you must install AC power at every location, that could be a lot of money, inspections, and the need of a licensed electrician. It all adds up. To help you along, see if you can connect to the nearest router. Most large buildings have a data room on every floor. Look at this as a haven for your equipment. If you don't have this luxury, then you may have to install the routers and cabling management at a central location with power available. Make sure to look at every floor. They may all be different causing roadblocks.

   d) Backhaul may cost you extra money. Most businesses and building have their connection to the internet so that you may be able to VPN, but is the bandwidth they have enough? Will they have to upgrade from that old T1 to something a little more modern? You may be the one who will break the news to them. It may add to your OpEx or theirs.

   e) If there are NOC services for monitoring, troubleshooting, or control, then someone must work out the SLA, Service Level Agreement. So, they need support 24/7 or just five days a week during business hours? It matters when someone must pay for people to be working or on call. If it's an automated service, then it may not be so bad if everything works the way it should.

   f) Try to think of what we missed, the unknowns, by asking the people who will rely on this. See what they hope to get out of this as far as a service or control.

5) Spectrum for building maintenance could include a few options.

   a) Wi-Fi for you to get smartphone data helps.

   b) For emergency workers, some buildings still have a 2way system to allow the workers to communicate real time, but with the coverage for cell phones and smartphones improving and with buildings adding small cells, this is starting to go away.

c) For maintenance, they may want to add an IOT system, see IOT Systems above, because then the gauges, equipment, and other devices could be monitored remotely, maybe from the NOC somewhere which could alert you to a problem before tenant does. It's an option that you should consider. However, I recommend having a design team or a radio shop design it. They need to understand your needs first. Use this guide as a foundation.

Entertainment, Stadium, Large Venue

1) Build your business case around the questions below. Then you can guide yourself through the RFP process. You may want to release an RFQ to get more ideas. The RFQ will get you other opinions on what to install, ideas to offer, and if you ask about applications, then you will get a completely different perspective.

2) Coverage is obvious, the venue. It is self-contained and yet there is so much to consider. Do you need to cover the parking lot? The streets? Why ask this? Follow along.

    a) Indoor coverage is the main concern. The carriers know that they should cover more than the venue. The owners of the venue want to cover the crowd inside the venue. Coverage inside the venue is key, but there is more to it than just the crowd. Remember that the vending stations all need to have good coverage, and they should be easy to cover. While the loading is not the issue here, they do need to be covered.

    b) Outdoor coverage is required. The parking lots, maybe places where people congregate like a kid's area. These places need coverage just like the indoor area needs it.

    c) Backhaul is a major concern. It's more than just throwing up hotspots. Your group needs to get the wireless data to a data room and out to the internet. I have bid on many NFL stadiums here in the US, and all want extensive Wi-Fi coverage. They wanted to have a massive backhaul and cover as many users as possible. They don't want to see any cables from the hotspots to the data room. It is a real issue running miles of conduit everywhere.

    d) They also wanted to offer their apps for the customer. It all works well, but it all takes time, and a massive amount of Wi-Fi access points to serve them well. Now they know. At first, they didn't understand how many it would take because they only looked at coverage, not loading. There is more to look at than coverage. If they want solid video for the apps, then you need to load balance the network and have the coverage solid enough to provide video to each user.

3) Goals of the venue owner usually are all about the experience. Today they know what it takes to give the spectator and great experience. It takes great coverage. It takes some apps that are only available in that stadium, like instant replay or access to the big monitor and maybe a venue specific coupon. They know it takes a quality customer experience for them to be satisfied.

a) When working with the customer, you may try to tell them all details, maybe make suggestions to improve. If they are in the bidding process, then chances are it's too late, they are only looking at money. I have worked through this before, and they do many construction projects, and when it gets to the bidding process, they know that most bidders are going to add features to add cost. The knee-jerk reaction is to push back and say this is what we want.

b) Understand the goals. Coverage, loading, apps, video, uplink, downlink, and anything else the customer may want.

c) Testing matters, set goals for the final test and sign off.

d) Set your goals up front, define them with clarity, if possible. Work with 1 or 2 consultants that can guide you to the proper outcome. All vendors will have great ideas, but if they don't align with your vision, then it's going to change what you want. The goal should be a quality experience for the end user. You and your customer determine that goal.

e) As a venue owner, think more about loading over coverage. You know you need to cover the venue, and you know where the dead spots will be. Obviously, you want to improve the system as people complain. However, if people have coverage but they can't get access then you have real problems. Think about how many users can get access at one time. Access points have limited connections, so they may get overloaded if you have too few. They can only connect to so many people at one time.

f) If you own the venue, are you going to provide support for the carriers? They will need space, and you want to have good coverage for the big 4 and whoever the local carrier is. They will be putting money into your stadium. What do you expect from them financially? I would not expect too much because they are doing you a favor by adding coverage in your stadium. Look at them as a teammate, not a customer. Work together for backhaul, mounting locations inside and outside the venue, and for general access. Share information as much as you can.

4) Budgets are going to include the system you want to have. Let's review.

   a) What equipment are you planning to use? Think of the type of equipment, the features, the performance expectations that you have then maybe align with several OEMs. Many use cheaper equipment to save money but disappointed when it doesn't work reliably or can't handle the traffic or fail often. Research the manufacturer and the model, remember they make different classes of equipment.

   b) Where will you mount the radios? Overhead is obvious, but there are so many hot spots that some stadiums put them under the seats. Location matters, you may need lifts, ladders, or special equipment. If you put them under the seats, then do you need to replace the seats or build special boxes? It begins to add up in your budget.

c) What apps will you include? While we are looking at deployment, the budget may encompass so much more, like developing new apps. More money is needed. I assume the overall budget covers this, consider the end to end view. Video draws more bandwidth and will be an issue. Loading the system more than usual.

d) The data room and the backhaul and the data equipment will be part of this and they

e) Once the system is built, and you get feedback, usually in the form of complaints. Then you can learn from it and improve. Now you need to some money to make improvements. Maybe there is a spot you chose not to cover because you didn't think people would use their smartphones there, but they do.

f) Take the feedback and improve. Reserve a budget for improvements.

5) Spectrum for a venue used to mean two things, DAS and Wi-Fi. We will have more options soon. Let me break it down below.

   a) Wi-Fi is still the standby. It may get some help from LTE-U as the carriers have shown support for this format in the unlicensed 5.8GHz band. Either way, it will be an easy thing to deploy. Easy doesn't mean "no planning or engineering." Easy means the spectrum is there and ready to use. The team still needs to plan for users, channels, and coverage.

   b) Carrier spectrum can still be used. It is going to depend on the venue as to which carriers will pay to install. DAS systems should be digital now, or the small cells should be deployed but what I have seen is that the carriers like to a macro eNodeB there to handle the heavy load. So far, larger venues have been overloading the small cells. I am not sure when the small cell model will be able to handle many users and 3 or more carrier aggregation. Someday soon I hope.

   c) The CBRS is another option. I don't know if this will be able to handle the proposed spectrum or if it will work with the existing carrier spectrum and Wi-Fi. So far, I don't see it as a viable play until more devices have this spectrum added. However, it would work well for special purposes, like for specialized equipment that you want to be on a lightly licensed band and not on the carrier's spectrum or open Wi-Fi spectrum. If you expect extreme loading, then you don't want on a shared network, especially one where you pay for data. I see it serving a different purpose than the standard crowd pleaser until devices carry the CBRS radios.

   d) Another special purpose band could be anything in the mmwave spectrum. It could be used to present virtual images of other games or entertainment. It may be something that the stadiums, with big budgets, could use to add value by having 3D imaging of other events. Could you imagine? Backhaul using mmwave has potential because of the massive spectrum.

Smart City

1) Build your business case around your needs. Think of who will use it. Most cities think that they can put everyone on one system, but that's not practical.

    a) The thing about a smart city is that there is more to it than wireless. Smart cities are about communication, efficiently and everywhere, but also about the energy savings and the lighting and the efficiency that will save money.

    b) The residents want a well-connected city. A city wants to run as cost effectively as possible and safety.

    c) The government wants a city with low costs for power and communications and a safe city.

    d) Tourists want a well-connected, (Wi-Fi and carrier coverage), and safe city. They also want to get around easy and find places to go, so apps that help them do that will give the city a good reputation and spreads a lot of good will.

    e) A smart city to be run smart, costs should be kept to a minimum for not only communications but for electrical and safety. It all adds up. If you are billing each department, from electric to water to gas to garbage, then think of each source of revenue. Think of the tourism and the residents because they will be paying the bills.

    f) What about the buildings? Most smart cities don't care about the buildings unless they are city buildings. Then they want to save costs. If you can make the buildings, say museums, part of the smart city rollout, it would help shine the light on what the features are of the city as well as provide cost savings for operations. Cities may do this because it could use grant money, a key to rolling out smart cities would be a great way to use it.

2) Coverage would be the city or area that they want to cover. I should be honest; most cities would never spend much money on a public wireless network. In fact, they would put it in, but they would not want to spend the money to maintain something like that. There are exceptions, like New York City where they put Wi-Fi hotspots around the city for the tourists and locals. That is where the public/private partnership comes into play. I'll discuss more on that below. The coverage for a city may seem obvious, but I've built networks where the cities cover not only city limits but the outskirts of town and other places nearby to work partnerships with neighboring communities and municipalities. Don't just think about city limits, think about partners. In this case, though you may want to read meters, turn lamps on and off, control vehicles throughout the city, monitor traffic lights or even sewer levels.

3) The goal of this coverage needs to be thought deeply about because you need to know what the use case is. Is it for video and security? Is it for the residents to have internet access? Is it for police and fire workers to have internet access? Is it for the workers to have network access? If you are building it for the city, think about how to maintain it as well as coverage.

The 5G Deployment Plan Handbook

The functionality may be for meters today, or lampposts, but what do you want to add to the system? What could be a potential target for a year?

4) The budget will be determined by what you use it for. The key to smart cities is to provide services and to save costs.

   a) The coverage will determine the budget. The coverage will not be solid in most cases. Make sure you figure it out what the needs are for the use. The thing is when the smart city planners start looking at the budget they will look beyond the wireless CapEx and into the OpEx. They will take the maintenance and monthly costs of the service into account. They will also look at what the electric bills will be. Who will be looking at the data? You could manage the system in a central location, or the system may need to send the data to different NOCs throughout the city. If you plan to cover multiple types of equipment, like from water levels to gas pressure levels to traffic lights, then you need to make sure each group can grab some of your data. Several cities are even thinking of garbage levels in waste bins so they can pick it up before it runs over. These are all target goals for coverage and send out data.

   b) Is it a prevailing wage area? Is it union required area? If you think it doesn't matter, let me tell you a story. I was working in Philly, coverage testing. To do this my partner and I had a bucket truck to test. The rules are the engineer could use a radio and a laptop but no tools. So, before we went into Philly, we put all the tools in the truck box and locked them up. We were down an ally, on a back street, behind some abandoned buildings. Sure enough, a black Lincoln pulls up, and the guy jumps out and starts yelling at us to see our union cards. My partner was ready for this and explained that we're engineers, not workers. He had the paperwork and our business cards. This guy didn't care; he started to call everyone, including the city. He got all the answers that we gave him, but that wasn't good enough, he called the union halls to make sure there was not work being done on that street. Then he watched us for the next 3 hours or so from his car. We went from street to street to test, and he followed us most of the day, making sure we didn't touch any tools.

   c) Are you providing outdoor coverage only or venues or streets? What about smart buildings? Traffic lights? What apps are you adding? What will be the use, video? Will you control the traffic lights or just monitor the traffic or do both? Will you turn the street lamps on and off? Are you adding free Wi-Fi for tourists and residents?

5) Spectrum for smart cities can vary based on what the goal is.

   a) I covered public safety in another section, but you may want to be sure that your emergency responders have high-speed internet throughout the city, the thing you need to look at is whether it's dedicated to them or if they will share a public network. What data will they be passing and what encryption will they be passing. If it is for data only, then it may not be mission critical. The only time it may be an issue of there is an emergency in an area where there are a lot of spectators and reports that are sharing the public Wi-Fi. The same happens to cell sites during an emergency; they get overloaded causing them to go down or drop calls.

b) Internet access for the citizens and tourist is one of the most obvious uses of Wi-Fi. Maybe LTE-U will start to ramp up but the way I see it, this is a great service. No one looks at Wi-Fi to have great coverage, but rather hotspots. So, this is something that you could strategically place on city-owned poles or kiosks. Always a good idea to make the residents and tourists happy.

c) Monitor and control traffic lights could take a licensed spectrum, low latency. It should be an IOT play, maybe in the 900MHz spectrum. Make sure you have the timers in there for backup.

d) Video will need high spectrum, like mmwave. Many cities have used Wi-Fi on a dedicated network. It worked well in the past when properly engineered. I would not just throw up a network, plan, and engineer.

e) Parking meters and garage space monitoring would be a great IOT play with low bandwidth. However, chances are you may have Wi-Fi in those areas, or maybe, in a parking garage you could have a video application. You may piggyback on of another system. It should all go back to a NOC to monitor. The deal with parking meters is that they may take a credit card which would need to be approved quickly or a remote need approval. Think through the application and latency before just jumping into an existing system.

f) If you are going to capture data to do analytics, then think of IOT. What are you going to monitor and control? Traffic lights? Video cameras? All of this is something that you may want to add to your smart city network. There's a variety of spectrum options. Just like reading meters, like parking meters and monitoring the parking garages. The video would add safety for the city and a way to track criminals pursued throughout the city.

Construction vehicles and sites

1) Build your business case around what the vehicle's purpose will be. Currently tested by Komatsu and Caterpillar, http://www.miningglobal.com/machinery/947/Top-Driverless-Trucks-in-the-Mining-Industry-Today-Plus-Future-Concepts for mining operations. They intend to have more autonomous trucks available to save on labor and efficiency. It's amazing, but it is happening with GPS and radar. Live video and sensors for the vehicles are the key issues. IOT is working hard to make mining efficient and safe. If someone can keep an eye on the trucks remotely and get an alarm when they need care or fuel, then they can work until they need to stop.
   a. There could be a business case for the workers to have cell and internet access while working. It could be 5G, LTE, or even push to talk. If it's 5G, then the business case becomes a little more concentrated to the workers as well as the vehicles.
2) Coverage of a vehicle like this is very specific. You think it would be easy to cover a work area. It is not so easy to provide coverage to these areas.

The 5G Deployment Plan Handbook

  a. Mining operations are a specific area. If it's a new mine, then you need to build the coverage up around and inside the mine. Underground coverage can always be a challenge. While this looks easy, the backhaul becomes an issue. It is no easy task to build this out. It takes a lot of planning to work things out.
  b. Construction sites are not always permanent. The covering of a temporary area needs should be thought out. Antenna placement is crucial for the workers. They need to have a coverage are that is solid and reliable, even though it is temporary. The radio and backhaul need the planning to meet the requirements.
  c. Oil operations are a very defined area, but mounting assets could be limited. While coverage is specific, mounting assets may cause you headaches.
3) Goals for this operation are very specific. Provide coverage to the designated area, but you need a system dense enough to cover the works as well?
  a. Vehicle coverage for autonomous vehicles will need to have good coverage, If monitoring video, then it will need a lot of bandwidth. If they want to have full control, then latency will need to be very low. No matter what, it needs to cover well. Above ground or below.
  b. If you plan to offer the workers coverage, will it be mission critical? Adding a whole new layer to what you're trying to achieve. If emergency coverage matters, then the goal goes from good enough to must have.
4) Budgets are going to be determined by what the goal is. The coverage based on the goal you set. Your CapEx will vary based on what you need to cover. In the past, I have built these systems using wood poles and monopole that line a road. If it's underground, then think about where to mount the radios in a confined space. Think about any hazards, like dangerous gas that may inhibit construction or could make the radio dangerous in the future. It adds money to deploy more sites. The backhaul could be an issue. Are you able to run fiber to all the locations or will you put in a wireless backhaul? Can you get away with an unlicensed multipoint backhaul? The budget has many variables. Maybe you can even build your core and tie into a carrier for extended coverage if that's the goal. LTE cores for small systems are becoming more and more affordable.
5) Spectrum for this application is all over the place, let's review.
  a. Unlicensed spectrum may work well in an isolated are where you can put in plenty of hotspots. LTE-U will offer more than Wi-Fi did as far as voice and devices and connections to the vehicles. Low latency will also really help if you have your core. The ISM band may work for you.
  b. Licensed spectrum will cost you more money, but maybe you don't need broadband. If you the goal is to control vehicles, then low bandwidth low latency spectrum will work fine.
  c. Leasing licensed spectrum from a carrier is an option if you are in a remote area. If it is part a carrier's nationwide license, then it's worth pursuing if it's a large project. Then you could use their core and become an MVNO.
  d. The CBRS, 3.5GHz band, may work for your system. If it's an isolated area, then you can get a lightly licensed band to work for you, and devices, then go for it.
  e. If you need the massive bandwidth, maybe you could use mmwave. Working in a small area, like a construction site, where you would provide a lot of bandwidth. I don't see

this as a viable solution, but it's an option you could have plenty of spectrum for a limited cost.

Renewable Energy

1) The business case for renewable energy is slightly different than the others. It could be a pure IOT play at houses. It could be a wind turbine out in a remote area, or it could be a network of solar panels in a suburban neighborhood.
2) Cover what you need to monitor or control.
    a. In a neighborhood, you might want to monitor a group of homeowners' solar panels or wind turbines in a neighborhood. Maybe a carrier's network or a wireless ISP or even connect it to the home owners network.
    b. For remote solar farms and wind turbines, you may need to have your backhaul. Many wind farms connect to a local network. They could then use the carrier's network to connect to a remote NOC. They could create their local wireless network, usually Wi-Fi but it could be LTE-U, and then use one modem to connect back into a larger 5G network. Most of the time these networks are used to monitor the turbines for alarms.
3) Set goals on how the devices will be deployed. Large wind turbines or solar farms need their network designed. In a rural neighborhood where you monitor homeowners' panels or turbines, then it's more of an IOT play.
4) The goal will determine budgets.
    a. For remote systems then you could build your local network all to be filtered back through a backhaul that could be fiber or wireless. Meaning that the CapEx using a 5G spectrum might not be so high, but the OpEx for the backhaul will run into some money each month. Plan it out and get quotes but be open-minded. I don't think you need broadband for this system so be as cost effective as you can to monitor these systems.
    b. For neighborhood systems, you could see if homeowners will let you tap into their Wi-Fi which should be free. Or you could you use a carrier and tap in using the IOT pricing. Maybe, you could connect to the SIGFOX system for coverage. All the same, the OpEx will cost more money than the CapEx.
5) Spectrum varies.
    a. Suburban networks may use Wi-Fi if they can tap into a home network or they could use a provider, like a carrier or someone like SigFox with a full IOT play. Wi-Fi or 900MHz. I don't see anyone building their network for an application like this, but I do see them jumping on an IOT network.
    b. Rural networks, like for the solar farms and wind turbine farms, need to use something to connect all the turbines or panels. If they want to connect each one individually, they could use a carrier that has coverage or someone like SigFox if there is coverage. If there is not enough coverage, then choose another provider or build a system to connect locally. Local coverage could use Wi-Fi or LTE-U which is easy and unlicensed. Another option is the 915MHz which falls into the ISM band and is unlicensed. An ideal band for short range, low bandwidth spectrum.

Gaming

1) Build your business case for gaming, but as you know, the gaming sales surpassed movie sales. That's right! Video games and online gaming are a huge industry. I brought up the entertainment aspect earlier, but the gaming industry is more than entertainment, it is an industry that will rely on massive broadband. It is something that deserves to have its breakout session. There may be places you can get to play among your peers. Look at the Pokémon Go sensation where people would see the Pokémon out in public. How cool was that? Now with broadband everywhere, you could do so much more.

2) Coverage could be just a designated area for a group of people to game together. There could be two scenarios.

    o Gaming groups. They could all be in one area with a massive wireless connection that could connect everyone at a very high bandwidth. That would be amazing. Instead of a quarter acre reserved for paintball, you could have a quarter of an acre reserved for a group of gamers going out and shooting avatars that could be anywhere. How amazing would it be that they could use their glasses to see what is live and a projected character created in virtual reality? Gaming in VR could be the next generation of the laser tag you see today.

    o In your home for those massive bandwidth needs in your home. It could take more than Wi-Fi spectrum to do the things that they want. Low latency and very hi bandwidth. Both are very demanding.

3) The goal is to have the very low latency and the very high bandwidth. Virtual reality will make the gaming industry go to new levels. It should make it a group participation more of a realistic goal. Several people using massive spectrum in an area for a game. So, make sure you have it ready to go. You also need to have servers near you that can run the programs. Your server will be the cloud server for the applications. The coverage area will be in a designated area. A specific area.

4) The budgets are constructed around the result. If you're covering a small area for gaming, then plan on several access points and a fiber connection between all of them with a server room somewhere nearby. If it's in a home, one access point may work, but two would be better.

5) Spectrum could be Wi-Fi or LTE-U, but I believe that the spectrum of choice will be mmwave. It should give access to a large amount of bandwidth and low latency. The band will allow enough spectrum for massive bandwidth that virtual reality can happen and the gaming functions. One thing you need is to maximize spectrum and minimize latency. The answer, mmwave, large swaths of bandwidth designated for a very small area. The fact is that mmwave could be an amazing solution for the gamers around the world. They could have ten people in a room all sharing the virtual reality goggles and glove to make the gaming experience even more real. We all know that virtual reality will be amazing, gaming will work out the kinks and make it improve and show us the wireless flaws. Go gamers!

Other – what will your business plan look like?

It's time for you to look at what you may want to do with 5G. Is it a broadband system? Low latency? IOT? Here is a model I have worked with that may help you move ahead with your system moving forward.

1) Build your business case around the initial idea.

   a. _____

   b. _____

2) Coverage based on what or who you plan to cover.

   a. _____

   b. _____

3) Goal setting is very important to help you identify what you intend to do for your customer or company or yourself.

   a. _____

   b. _____

4) Budgets can be set and adjusted now that you have a clearer vision of what you intend to do.

   a. _____

   b. _____

5) Spectrum may be something you can determine or you may be forced to use a carrier or IOT service provider. You have all the building blocks in place, now determine if you can build your own system with licensed spectrum, unlicensed, 3.5GHz, or mmwave that is lightly licensed. What can you afford and what can you get?

   a. _____

   b. _____

IOT

What is NB IOT and how will we use it?

Are you curious about IOT? Do you know the different flavors of IOT? Do you wonder why they use IOT and 5G interchangeably? Who would use this technology? Read on wireless tech fan if you want to find out.

Isn't IOT just the internet of things? Do we need so many variations? YES! There are different flavors of IOT that are available. Each one has a different use case. The one thing that you learned about 5G is that it will be made up of HetNet. If you're building out a network, look at NB-IOT as a very powerful tool in your war chest. Any of us could install a smart thermostat or light switch in our homes, right? While that is cool, it's very limited, and anyone can do it. For IOT you will want to build a business plan around deploying the network to control thousands of meters for a utility, remote devices, alarms, track equipment, and more. A way for you to break the network down to each specialty item. For low bandwidth, high battery life remote devices, this is perfect. Maybe open and close doors remotely in a building or track where all the equipment is in real time, like for a corporate building or a hospital. Pretty cool, right?

What is NB-IOT?

It is Narrowband Internet of Things which is, according to Wikipedia, " *is a Low Power Wide Area Network (LPWAN) radio technology standard that has been developed to enable a wide range of devices and services to be connected using cellular telecommunications bands. NB-IoT is a narrowband radio technology designed for the Internet of Things(IoT), and is one of a range of Mobile IoT (MIoT) technologies standardized by the 3rd Generation Partnership Project (3GPP)."* So basically, it is a technology used for machines that have low data needs and don't need to be connected all the time. The connection issue is so that the battery life I extended, hopefully, batteries will last ten years. By that time, you may replace the unit altogether because of the advances in technology.

Why do people use 5G and IOT interchangeably I do? It doesn't make sense to me mainly because I look at 5G as the network and I look at IOT as a service. Now, IOT will be a big reason for deploying 5G. However, there is also NB IOT for applications in IOT devices. The application is pretty specific. Let's look at how they will be used and who will use them.

Let's look at NB-IOT which I believe is above 3GPP release 13. Yes, 3GPP is developing this along with the IOT technologies, so it's not a fly by night format. It is a low-power wide area network, LPWAN, format. It means just what it says, low power RF reaching many devices over a wide area. Its major focus is to reach low power devices, something that may need the battery to last over ten years. Limited bandwidth using 250 Kbps up and 250Kbps down. Very low bandwidth. Not for video, or any high bandwidth application. A meter could report, "I'm reading 35 degrees Fahrenheit" or for the system to tell a relay to close or open. These are short data bursts that don't require much data to send a simple machine language command. It is half duplex, meaning it will talk and then listen. The antennas are very simple, one transmit, and one receive, SISO which is single in single out. Transmit power is very low, 23dBm, around 200 milliwatts of power.

It is very limited, but how can it help I the IOT deployments? Who would use this? In large buildings, there could be a need for something like this to track equipment or open and close doors. It would be low latency and quick access because it would be a dedicated network. Most likely be an add-on to an existing network. It would be specifically to contact devices that you may only need to poll once a week or so. It'ss meant to communicate with objects that can give you a simple response or one that you could send a simple command. If you have a large outdoor network, then you could use this to extend it or talk to devices that run on the battery. Maybe even to use for security to send out an alert if it is tripped in a remote area. Look at this as another tool in your HetNet arsenal. A network that you could deploy cost-effectively to communicate with devices that are very remote or don't have access to power.

There are several possibilities, like remote fence alarms, meters, equipment tracking, animal tracking, and more. An extension of the larger network.

The spectrum is an issue for me here in the USA because there is not dedicated spectrum that I found. It looks like anyone who may have narrowband spectrum may be able to use this. I did read that GSM spectrum would work. I don't see much of that spectrum nationwide, but I am not sure if a nationwide deployment would be necessary. It looks like 200Khz of bandwidth would work for something like this. Would the GSM bands be re-farmed to run something like this? It could be, why not? What an opportunity to build something for IOT only.

While I said, this is part of the HetNet. There are attempts to build large networks to cover these specialty IOT circumstances. When you think of all the devices and systems that can benefit from a network like this. It could be more than just remote devices. Think about agriculture, metering, lighting control, smart city control and monitoring, and industrial equipment. See the value of a system that could keep these devices connected with extremely long battery life? Remember that these devices only need small data bytes to communicate. What a great opportunity for a new specialized network to be built.

All this when you thought small data networks were history. We need to connect everything, but we need to be smart about it. It means that we can build a better strategy for these specialized networks and hopefully we won't overcharge customers like the carriers intend to. It makes better business sense to build a cost-effective system for these systems where they don't want the bandwidth or constant connections. They want to have a connection for a few dollars a month. Here's a great business plan!

Resources:

https://en.wikipedia.org/wiki/NarrowBand_IOT

http://www.samsung.com/global/business-images/insights/2016/IoT-Whitepaper-0.pdf

http://www.huawei.com/minisite/4-5g/img/NB-IOT.pdf

http://resources.alcatel-lucent.com/asset/200178

https://networks.nokia.com/innovation/iot

http://www.ericsson.com/res/docs/whitepapers/wp_iot.pdf

Glossary - Naming Overview (Abbreviations and Acronyms)

All the names and acronyms can be confusing. It adds a great deal of confusion to the deployment teams as well as some of the carriers. So, let's start by helping you know what I am talking about to avoid as much confusion as possible.

- **3GPP** – Per the website, http://www.3gpp.org/about-3gpp, "The 3rd Generation Partnership Project (3GPP) unites [Seven] telecommunications standard development organizations (ARIB, ATIS, CCSA, ETSI, TSDSI, TTA, TTC), known as "Organizational Partners" and provides their members with a stable environment to produce the Reports and Specifications that define 3GPP technologies." It started with 3G, and they kept working to set standards moving forward. You can also learn more on Wikipedia, https://en.wikipedia.org/wiki/3GPP if you want to learn more about what they do.

- **Het Net** – A Heterogeneous Network is a network that includes multiple transmissions using multiple ways to deliver data. Per Wikipedia, "A **heterogeneous network** is a network connecting computers and other devices with different operating systems and/or protocols. For example, local area networks (LANs) that connect Microsoft and Linux based personal computers with Apple Macintosh computers are heterogeneous. The word **heterogeneous network** is also used in networks using different access technologies. For example, a wireless network which provides a service through a wireless LAN and is able to maintain the service when switching to a cellular network is called a wireless heterogeneous network." Another good resource is http://www.3gpp.org/hetnet if you're interested. What we are looking at for the carrier is how to deliver the wireless data/voice using a macro, small cells, indoor, and even Wi-Fi to the customer. This network is going to deliver the data using multiple devices, technologies, and frequencies.

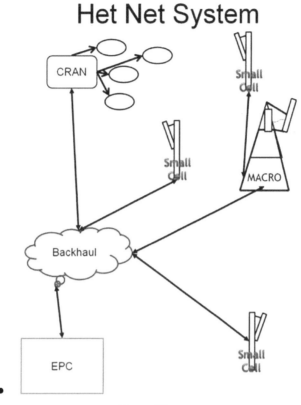

Figure 21

- **BBU** – this is for Base Band Unit. This unit that interfaces between the backhaul and the RRH. It will process the incoming data to be sent to the RRH for transmission to the end user. This unit can run on AC or DC. Generally, with a small cell, it is part of the small cell, not a separate unit. In most configurations, this would be a standalone unit. It's the case for a macro site or a CRAN site.

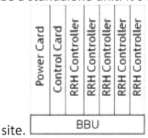

Figure 22

- **hBBU** – This would be a host BBU, also called a BBU hotel, where a BBU would connect to remote radio heads at another location.

- **RRH** – Remote Radio Head – this is the RF unit that is not attached to a BBU. It could be on a tower where the BBU is on the ground in a cabinet, and the RRH is up on the tower with the antennas. They split them to save on RF loss in the cables and to save on real estate on the ground. However, now you are using more real estate on the tower and increasing the loading on the tower. In a DAS setting, indoor or outdoor, it would be the RF unit, (radio) that would be mounted with behind the antenna or attached to the antenna.

The 5G Deployment Plan Handbook

Figure 23

- **Macro site** – this is a full-blown site, it could be on a tower or a building top or even a stealth site. It consists of fiber backhaul, battery backup, a large BBU, several RRHs and several antennas. It would make up several sectors, usually pointing in many directions. It would also have a router that would connect it back to the core, which is a fiber backhaul, but it could be wireless. It isn't copper, but it could be.

- **BTS** – Base Transceiver Station – the RF base station.

- **eNodeB** – Evolved Node B from the E-UTRAN This included the air interface, the BTS, and the interface to the EPC, the evolved packet core. Normally all the equipment at the cell site.

- **E-UTRAN** – evolved universal terrestrial Radio access network. It would be all the cell sites in the network.

- **UE** – User Equipment, this is the cell phone, smartphone, tablet, or any end user wireless device. The user would be the person carrying the device using the wireless connection.

- **Small Cell:** The small cell where the unit has everything included, except maybe the router, but the BBU, (broadband unit) and the Radio Head are all in one unit. Small cells radiate less than 5 watts outside and less than 1 watt indoors.

- **Femto Cell:** A small cell. It would be very small, and one that normally would self-configure and you could plug into any internet access for backhaul. It may be something to put in a home or a small business to improve coverage. It would cover a small area to provide better signal and offloading. Usually just for a few UE connections, maybe up to 10. Most of the time it just connects to the internet, like someone's cable modem.

- **Pico Cell:** This is slightly bigger than the Femto, usually for a mid-sized business, small bus station, or a smaller public area to connect maybe 10 to 100 connections at any given time.

- **Micro Cell:** This is usually a bigger unit that can handle medium stadiums or a train station or airport. This term is not used so much anymore because they just call it a small cell.

- **Metro Cell:** This term was used for larger outdoor metro areas where the loading could be greater than 100 users at any given time. But let's face it, they are commonly called small cells.

- **Indoor small cell:** It is a small cell that is mounted inside a building to provide coverage and /or offloading of the Macro site. This unit would not be weather proofed.

- **Outdoor Small Cell:** It is a small cell mounted outdoors to help with coverage and offloading of the Macro site. This unit would be weatherproofed, and it may need a sun shield.

- **CRAN** – this is for Centralized RAN, where you would have the controller, a BBU, in one area and it would control several RRHs as part of a distributed radio system. This setup indoors or outdoors could be part of a DAS system. The RRH could be at the top of a pole, other parts of a building or a few miles away mounted on a light pole. The BBU may be in a different location than the RRH. So common in today's world. You know that on the towers they put the BBU on the ground, and the RRH is up on the tower with the antenna. So just imagine now that they put the BBU in a basement and the RRH's will be spread throughout the building. They may also have the BBU in a closet at a building and spread the RRH's all over town to get the RF where the people are, to distribute the radio heads and antennas. It could be part of a DAS system where they rely on the CPRI, (common public radio interface), to be connected to fiber for the "fronthaul" which is like the backhaul but to go from the BBU to the RRH, forward! Currently, there are several limitations which mostly should do with timing. They can only travel so far before they would time out. That limits distance now. The cloud may change that soon. When they locate many BBUs in a remote location for widely distributed RRHs, they call that a BBU hotel, a term that means that many BBUs for multiple locations are in one spot. The BBU will be farther and farther away from the radio heads.

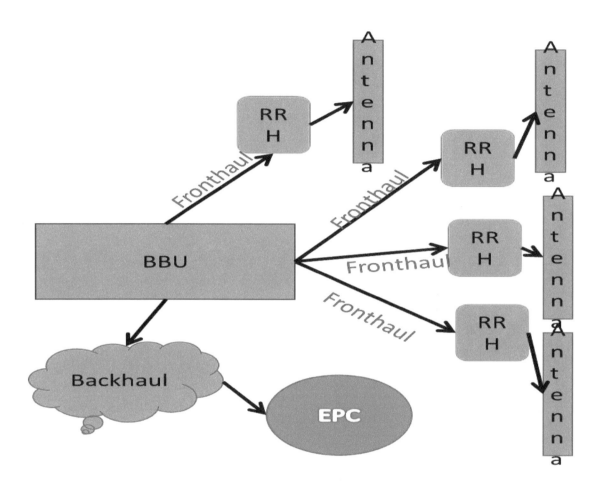

Figure 24

-
- **vRAN or cRAN** – this stand for virtualized RAN and cloud RAN. These are part of a CRAN but the vRAN and cRAN will, (in theory) can use standard networking protocols to communicate to between the BBU and the RRH. These systems are out there but not available at this writing, but they should be released very soon.
- **DSCS** - Distributed Small Cell System is where you would deploy small cells, (and maybe Wi-Fi) like you would with a DAS system. These would be standalone all-inclusive small cells with integrated antennas. They would be connected to the EPC with fiber or CAT5 or some type of wireless backhaul.
- **DAS** – Distributed antenna system. This could be 1) an eNodeB feeds the RF distribution system, and RF cables run from the central area to each antenna remotely which is like most **iDAS**, (indoor DAS) systems or 2) there could be a central **hBBU** (host BBU) that uses fiber or CAT 5 to feed

several radio heads with antennas which is common in **oDAS** (outdoor DAS). Together they form a solution for coverage. There are many more DAS solutions, but for LTE let's stick to these 2.

- **Mini-Macro** – this is a medium power eNodeB installed like a one sector macro site. It would be a single sector cell site, lower power, with a single antenna. There will be variations but a stand-alone, medium power cell site. Since it would resemble a smaller Macro site, it would have a larger cabinet to host the BBU and the router mounted at the same location as the RRH. The RRH would be up on the pole or tower or building with the antenna.

- **Backhaul** – the link from the BBU (or small cell) back to the EPC, (the core). It usually goes through a router and then to fiber, wireless, or a copper link. It could go through the Internet, or it could be a dedicated line or a VPN back to the RAN's core.

- **Fronthaul** – the link that connects the BBU to the RRH. It could be fiber, copper, or wireless. At the time of this writing, the latency is critical. There are limitations on distance and technology because if the signal arrives too late to the RRH, then you will have problems.

- **UE Relay** – this is an option used for backhaul where you would use the carrier's band for backhaul. Called UE Relay because it would have a UE, (User Equipment) like a phone as a receiver that would interface to the BBU or small cell to connect back to another macro site. It would be like a repeater, but the carrier would dedicate part of the band for backhaul. IF the carrier is willing to use their valuable spectrum for backhaul, this is a great option. It would depend on how much spare spectrum they have and how they want to use it.

- **EPC** – Evolved Packet Core is the core. Here are the brains are for the RAN to interface with the world! Here's where the 4G RAN will connect, via backhaul, for control and connectivity. Currently, in 2015, there are still regional cores but when the EPC and the cloud evolve there could be one core for the entire system utilizing the cloud as the core becomes totally virtual. The vEPC, virtual EPC, where the core will be in one location and communicate to the RAN via the cloud. Core functions will be in the cloud. Currently being developed, but the delays are still an issue. There are also other issues like multi-vendor RANs, where one core will need to control all OEM RANs.

- **PoE** – Power over Ethernet. Where the router or BBU or power brick supply power over a CAT5 or CAT6 cable, along with data, to the remote unit/radio/small cell. Helpful for the deployment of small cells and RRH and Wi-Fi units because of the power and data source, (router or BBU), can be in one area away from the remote unit, (RRH, Wi-Fi, or Small Cell). Only need that one connection for power and data, the CAT copper cable. Making installations easier when you connect a remote unit, and you don't need to run separate cables or have a local power source for the small cell.

- **OEM** – Original Equipment Manufacturer, this is the company that makes the equipment, that simple. For instance, in the world of small cells, it could be Alcatel-Lucent, Nokia, Ericsson, Samsung, SpiderCloud, Nirvana, and Airspan. There are many but let's go with these vendors now.

- **CPRI** – Common Public Radio Interface is used in a fronthaul situation where you have the BBU connected to the RRH. It is the common interface that is utilized by some OEMs. For more information, I have a link, http://www.cpri.info/spec.html which will give you more information on this interface.

- **OpEx** – Operating expenses are the expenses that the user must pay month after month. In most of this book, I will refer to the carriers, your customers because they are the ones that must pay the monthly expenses like rent, electric, and backhaul bills.

- **CapEx** – Capital expenditures are the upfront costs of the installation. In the case of the carriers, it is the site acquisition, engineering, installation, commissioning, and optimizations fees. I am sure there are more, but this is the one-time cost that your customer will pay to deploy each site.

- **RAT** – Radio Access Technology – this is used when covering several different types of technologies like CDMA, LTE, and Wi-Fi.

- **VoWiFi** – Voice over Wi-Fi is where the mobile phone can handle a voice phone call on a Wi-Fi carrier.

- **VoLTE** – Voice over LTE is where the mobile phone can handle a voice phone call over LTE.

- **QoE** – Quality of Experience is where the customer's experience of using the UE is rated. Voice quality and downloading experiences matter. If the customer has a connection, but they can't download anything or if they are using VoLTE and the quality of voice is horrible, then they will have a poor QoE. Get it? It's very important to the carriers because if they have dropped calls or they can't download a file, then this carrier sucks!

- **DOCSIS®** – this is a connection to a cable company using their direct connection for power and backhaul. DOCSIS stands for Data Over Cable Service Interface Specification.

- **CATV** – Cable TV acronym.

- **LTE** – Long Term Evolution.

- **LTE**-U – LTE in the Unlicensed spectrum, like ISM where Wi-Fi resides.

- **LTE**-V – LTE for Vehicles.

- **LTE**-A – LTE Advanced, faster LTE moving from 4G to 5G.

- **LTE**-B – LTE broadcast format for broadcast like live sporting events. While part of the 4G adoption it should grow with 5G broadband. Ericsson is really on top of this format. For more look at https://www.ericsson.com/res/thecompany/docs/publications/ericsson_review/2013/er-lte-broadcast.pdf

- **LTE**-4.5 – LTE in 4.5G, which is (in theory) faster than 4G but not quite 5G, made up by one of the OEMs.

- **LTE**-M – LTE machines, or LTE M2M or LTE for IOT. Now the 3GPP has CAT-M1 standards and CAT-NB1 which are both IOT standards.

- **LTE-FDD** – Long Term Evolution full division duplex. A carrier would have paired spectrum where the band is paired. One band is dedicated to the uplink, and the other is dedicated to the downlink.

- **LTE-TDD** – Long Term Evolution time division duplex. The carrier is all in one band, and it will use the same spectrum for the uplink and downlink. You don't need "paired spectrum" because it is all done in one carrier. There is a great explanation here.

- **Link Budget** – This is the accounting of all gains and losses for a wireless link. Gains would be power out of the transmitter, antenna gain on both sides, and receiver sensitivity. Losses would be cable or waveguide loss, connector loss, air loss for the free space distance between the links. The link budget for a fiber link would be the loss through the fiber and the distance between the fiber links.

- **eICIC** – enhanced Inter-cell interference coordination which is the interference on the same channel that a cell can have on a neighboring cell or sector. Learn more here.
- **QoS** – Quality of Service
- **BOM** – Bill of Materials, this is the list of materials for the job. There will be several BOMs, one for the system, and one for the cluster, and one for each site. If you drill down enough, there is one for each eNodeB and each assembly, even each cable.
- **SOW** – Scope of Work, this is the document that defines the scope of work. There are several, a high level of the system, then for each site, there should be a scope defined for the work to be done at a site. When working small cells, they may have one generic scope, and it may be up to the installer to figure it all out. Each job function will have a scope with milestones and expectations in it. The RF Design team will have a scope of work like the commissioning and integration teams. Everyone needs a guideline to follow and a defined end to get paid.
- **KPI** – Key Performance Indicators, these are the measurements made to get paid or at least to get graded.
- **IOT** – Internet of Things – machine to machine communications. Physical objects are talking to other physical objects, and possibly controlling them and communication to make specific changes on Internet-enabled devices.
- **MM-Wave** – MMW, millimeter waves are found in the EHF bands, usually anywhere from 30 to 200 Gigahertz, but we also include 24GHz and 28GHz in this because it is so close and it makes it easier to follow. Learn more at Wikipedia, link found here. A LOS signal.
- **IEEE 02.11ad** – part of the 802.11, the ad version refers to the 60GHz band and its design for operation. It could also mean WiGig.
- **WiGig** – part of the 802.11ad description with unlicensed 57GHz to 64GHz (which uses four 2.16GHz channels) in mind. It also has some extra security functions found in it. Uses OFDM to support up to 7Gbps of throughput. A single carrier can deliver 4.6Gbps of data. Can use AES.
- **AES** – Advanced Encryption Standard.
- **OFDM** – Orthogonal Frequency Division Multiplexing
- **ISM** – Industrial-scientific-medical band used for mostly license-free communications.
- **LOS** – Line of site
- **FOG computing** – this is where the computing becomes real time computing, done so close to the end user that the latency is extremely low because it only goes a short distance.
- **Latency** – The time it takes for a packet to leave its original location and reach its destination then return. Click here, http://whatis.techtarget.com/definition/latency for more detail.
- **MVNO** – this could be Multi-Vendor Network Operator or Mobile Virtual Network Operator.
- **WISP** – Wireless Internet Service Provider – this is someone who uses wireless to provide the internet to their customers.

- **IOT** – Internet of things, this is the term used for machine to machine language where the device is connected directly to the network, sometimes the internet. The device has limited or no way for the user to connect directly but only through a network connection.
- **PIM** – Passive Intermodulation which is a test, the problem is caused when RF mixes causing an intermod signal to pop up, find out more at https://www.anritsu.com/en-US/test-measurement/technologies/pim to learn more.
- **HIPAA** – The Health Insurance Portability and Accountability Act of 1996 which covers privacy in health care and covers many aspects of privacy for patients. For a summary, go to https://www.hhs.gov/sites/default/files/privacysummary.pdf to learn more.

A Note from Wade

Thank you for taking the time to learn more. You are awesome! If you are planning to do deployments, then you should understand the steps needed to deploy. I want to help you out with more than just the high-level posts I put out on www.wade4wireless.com but more detailed. It may not have everything, but it should be a good step to educating you in CRAN and small cell deployment. You deserve to understand the wireless deployment world! I am assuming that you know something about the business of wireless deployment and optimization. If you have more questions, you can always email me at wade4wireless@gmail.com, and I will help any way I can. If I don't respond, then email me again, I do get busy and sometimes overwhelmed.

I currently run TECHFECTA, a company that has four main goals:

1. Advise tech startups and investors of technology and market trends.
2. Consult companies on deployments and bringing new products to market.
3. Create technology reports.
4. Create reports on technology and deployments. I do the research, so you don't have to.

Over the last several years I have been focused on small cell deployment, indoors and outdoors, as well as macro deployments long before that. The larger Het Net deployments are starting. I see the upgrades on the existing towers as well as the deployment of the stand-alone sites, mostly CRAN and small cells but some mini macro, 10 to 20 watts, cell sites for "densification" of LTE systems to help offload macro sites. So, let's talk about that.

I have been in wireless for over 25 years. I worked in paging and 2-way in the late 1980s all the way to 1999. Then I got into the wireless internet craze, public safety, broadcast, and then into cellular. I worked as an engineer and as a tower climber, two jobs that went together quite well. I had many positions, from technician to engineer to PM to Vice President and probably a few more. Trust me, no matter what job I had; I often felt like the mere peon, or should I say "peed on," most of the time. I was willing to get my hands dirty, and not everyone likes that.

I have been on both sides, meaning I deployed in the field for years and then I have worked in business development and sales. I would say the best analyst out there is the one on the front line! When you are in the field, you see all the problems, and they filter back through the ranks depending on how much they slow the project down.

One of the great things about wireless is that there are so many industries to work in, like the internet, broadcast, public safety, and cellular. However, with the growth of LTE, I think that they are all going to merge. There is still a need for wireless backhaul. I don't know if you can get bored in this industry, but I do know that you can get burned out.

Don't forget to follow the wireless deployment blog at www.wade4wireless.com today! Make sure you follow the podcast on iTunes or Stitcher, so you stay up to date! I also have email, wade4wireless@gmail.com for you to reach out anytime.

Other Books by Wade

In case you were interested, I have several products out there.

- Tower Climbing: An Introduction
- Scope of Work Tutorial
- The Wireless Deployment Handbook: Small Cells, CRAN, and DAS edition

Extras

More business plan sheets:

It's time for you to look at what you may want to do with 5G. Is it a broadband system? Low latency? IOT? Here is a model I have worked with that may help you move ahead with your system moving forward.

1) Build your business case around the initial idea.

   a. _____

   b. _____

2) Coverage based on what or who you plan to cover.

   a. _____

   b. _____

3) Goal setting is very important to help you identify what you intend to do for your customer or company or yourself.

   a. _____

   b. _____

4) Budgets can be set and adjusted now that you have a clearer vision of what you intend to do.

   a. _____

   b. _____

5) Spectrum may be something you can determine or you may be forced to use a carrier or IOT service provider. You have all the building blocks in place, now determine if you can build your own system with licensed spectrum, unlicensed, 3.5GHz, or mmwave that is lightly licensed. What can you afford and what can you get?

   a. _____

   b. _____

Other – Write your business plan.

It's time for you to look at what you may want to do with 5G. Is it a broadband system? Low latency? IOT? Here is a model I have worked with that may help you move ahead with your system moving forward.

1) Build your business case around the initial idea.

    a. _____

    b. _____

2) Coverage based on what or who you plan to cover.

    a. _____

    b. _____

3) Goal setting is very important to help you identify what you intend to do for your customer or company or yourself.

    a. _____

    b. _____

4) Budgets can be set and adjusted now that you have a clearer vision of what you intend to do.

    a. _____

    b. _____

5) Spectrum may be something you can determine or you may be forced to use a carrier or IOT service provider. You have all the building blocks in place, now determine if you can build your own system with licensed spectrum, unlicensed, 3.5GHz, or mmwave that is lightly licensed. What can you afford and what can you get?

    a. _____

    b. _____

The 5G Deployment Plan Handbook

1) Build your business case around the initial idea.

   a. _____

   b. _____

2) Coverage based on what or who you plan to cover.

   a. _____

   b. _____

3) Goal setting is very important to help you identify what you intend to do for your customer or company or yourself.

   a. _____

   b. _____

4) Budgets can be set and adjusted now that you have a clearer vision of what you intend to do.

   a. _____

   b. _____

5) Spectrum may be something you can determine or you may be forced to use a carrier or IOT service provider. You have all the building blocks in place, now determine if you can build your own system with licensed spectrum, unlicensed, 3.5GHz, or mmwave that is lightly licensed. What can you afford and what can you get?

   a. _____

   b. _____

The 5G Deployment Plan Handbook

Scope of Work Outlines Cover Sheet

Cover Sheet

Customer name: _____

Offer date: _____   Offer number: _____

Customer Contact name: _____ Contact number: _____

Offer Name: _____

Site address: _____   Site Coordinates: _____

Site ID/reference: _____

Site details: _____
_____

Detailed Offer Description: _____
_____
_____
_____
_____
_____
_____

BOM

Part Name	Part Number	Price each	Quantity	Description

Price to Customer: _____

Assumptions: _____

Exceptions: _____

Scope of Work Details

Site Name: _____

Supplier work description: _____

_____

_____

_____

_____

_____

_____

_____

_____

_____

Entrance Criteria: _____

Customer will provide: _____

_____

_____

Options: _____

_____

_____

_____

Additional notes: _____

_____

_____

_____

The basics of 5G from a deployment perspective. This book is not covering the technical details of the radio equipment or chips, Qualcomm and Intel have done a pretty good job of covering that. Don't worry, Nokia, Ericsson, and Samsung will explain it to the carriers with great detail. The goal was to respond to all of your questions about the practical deployment of 5G in today's wireless networks. How can others outside of the carrier domain deploy? You wanted to know the foundation, well here it is. Business case summaries and all. Enjoy and deploy!

www.wade4wireless.com

www.TECHFECTA.com

CPSIA information can be obtained
at www.ICGtesting.com
Printed in the USA
LVHW070230030520
654894LV00005B/331